PROBLEM-SOLVING CASES IN
MICROSOFT® ACCESS™ AND EXCEL®

PROBLEM-SOLVING CASES IN MICROSOFT® ACCESS™ AND EXCEL®

Annual Tenth Edition

Ellen F. Monk

Joseph A. Brady

Gerard S. Cook

COURSE TECHNOLOGY
CENGAGE Learning·

Australia • Brazil • Japan • Korea • Mexico • Singapore • Spain • United Kingdom • United States

COURSE TECHNOLOGY
CENGAGE Learning·

**Problem-Solving Cases in Microsoft®
Access™ and Excel®, Annual Tenth Edition**
Ellen F. Monk, Joseph A. Brady,
Gerard S. Cook

Publisher: Joe Sabatino

Senior Acquisitions Editor: Charles
McCormick, Jr.

Senior Product Manager: Kate Mason

Development Editor: Dan Seiter

Marketing Director: Keri Witman

Marketing Manager: Adam Marsh

Senior Marketing Communications Manager:
Libby Shipp

Marketing Coordinator: Eileen Corcoran

Design Direction, Production Management,
and Composition: PreMediaGlobal

Media Editor: Chris Valentine

Cover Images: ©violetkaipa/Shutterstock,
©sdecoret/Shutterstock

Manufacturing Planner: Julio Esperas

For product information and technology assistance, contact us at
Cengage Learning Customer & Sales Support, 1-800-354-9706.
For permission to use material from this text or product,
submit all requests online at **cengage.com/permissions**.
Further permissions questions can be e-mailed to
permissionrequest@cengage.com.

Some of the product names and company names used in this book have been used for identification purposes only and may be trademarks or registered trademarks of their respective manufacturers and sellers.

Library of Congress Control Number: 2011945044

ISBN-13: 978-1-133-62979-5

ISBN-10: 1-133-62979-2

Course Technology
20 Channel Center Street
Boston, MA 02210
USA

Screenshots for this book were created using Microsoft Access and Excel®, and were used with permission from Microsoft.

Microsoft and the Office logo are either registered trademarks or trademarks of Microsoft Corporation in the United States and/or other countries. Course Technology, a part of Cengage Learning, is an independent entity from the Microsoft Corporation, and is not affiliated with Microsoft in any manner.

The programs in this book are for instructional purposes only. They have been tested with care, but are not guaranteed for any particular intent beyond educational purposes. The author and the publisher do not offer any warranties or representations, nor do they accept any liabilities with respect to the programs.

Course Technology, a part of Cengage Learning, reserves the right to revise this publication and make changes from time to time in its content without notice.

Cengage Learning is a leading provider of customized learning solutions with office locations around the globe, including Singapore, the United Kingdom, Australia, Mexico, Brazil, and Japan. Locate your local office at: **www.cengage.com/global**.

Cengage Learning products are represented in Canada by Nelson Education, Ltd.

To learn more about Course Technology, visit **www.cengage.com/coursetechnology**.

Purchase any of our products at your local college store or at our preferred online store, **www.cengagebrain.com**.

Printed in the United States of America
1 2 3 4 5 6 7 16 15 14 13 12

To my students, for their curiosity and hard work.
EFM

To all the world's problem solvers.
JAB

To Ellen and Joe, for inviting me to join a winning team,
and to my wife Rita, for her love and support.
GSC

BRIEF CONTENTS

For two decades, we have taught MIS courses at the university level. From the start, we wanted to use good computer-based case studies for the database and decision-support portions of our courses.

At first, we could not find a casebook that met our needs! This surprised us because we thought our requirements were not unreasonable. First, we wanted cases that asked students to think about real-world business situations. Second, we wanted cases that provided students with hands-on experience, using the kind of software that they had learned to use in their computer literacy courses—and that they would later use in business. Third, we wanted cases that would strengthen students' ability to analyze a problem, examine alternative solutions, and implement a solution using software. Undeterred by the lack of casebooks, we wrote our own, and Course Technology, part of Cengage Learning, published it.

This is the tenth casebook we have written for Course Technology. The cases are all new, and the tutorials are updated.

As with our prior casebooks, we include tutorials that prepare students for the cases, which are challenging but doable. The cases are organized to help students think about the logic of each case's business problem and then about how to use the software to solve the business problem. The cases fit well in an undergraduate MIS course, an MBA information systems course, or a computer science course devoted to business-oriented programming.

BOOK ORGANIZATION

The book is organized into seven parts:

- Database cases using Access
- Decision support cases using the Excel Scenario Manager
- Decision support cases using the Excel Solver
- A decision support case using basic Excel functionality
- Integration cases using Access and Excel
- Advanced Excel skills
- Presentation skills

Part 1 begins with two tutorials that prepare students for the Access case studies. Parts 2 and 3 each begin with a tutorial that prepares students for the Excel case studies. All four tutorials provide students with hands-on practice in using the software's more advanced features—the kind of support that other books about Access and Excel do not provide. Part 4 asks students to use Excel's basic functionality for decision support. Part 5 challenges students to use both Access and Excel to find a solution to a business problem. Part 6 is a tutorial that teaches advanced skills students might need to complete some of the Excel cases. Part 7 is a tutorial that hones students' skills in creating and delivering an oral presentation to business managers. The next sections explore these parts of the book in more depth.

Part 1: Database Cases Using Access

This section begins with two tutorials and then presents five case studies.

Tutorial A: Database Design

This tutorial helps students understand how to set up tables to create a database, without requiring students to learn formal analysis and design methods, such as data normalization.

Tutorial B: Microsoft Access

The second tutorial teaches students the more advanced features of Access queries and reports—features that students will need to know to complete the cases.

Cases 1–5

Five database cases follow Tutorials A and B. The students must use the Access database in each case to create forms, queries, and reports that help management. The first case is an easier "warm-up" case. The next four cases require more effort to design the database and implement the results.

Part 2: Decision Support Cases Using Excel Scenario Manager

This section has one tutorial and two decision support cases that require the use of the Excel Scenario Manager.

Tutorial C: Building a Decision Support System in Excel

This section begins with a tutorial that uses Excel to explain decision support and fundamental concepts of spreadsheet design. The case emphasizes the use of Scenario Manager to organize the output of multiple "what-if" scenarios.

Cases 6–7

Students can perform these two cases with or without Scenario Manager, although it is nicely suited to both cases. In each case, students must use Excel to model two or more solutions to a problem. Students then use the model outputs to identify and document the preferred solution in a memorandum. The instructor might also require students to summarize their solutions in oral presentations.

Part 3: Decision Support Cases Using Microsoft Excel Solver

This section has one tutorial and two decision support cases that require the use of Excel Solver.

Tutorial D: Building a Decision Support System Using Microsoft Excel Solver

This section begins with a tutorial for using Excel Solver, a powerful decision support tool for solving optimization problems.

Cases 8–9

Once again, students use Excel and the Solver tool in each case to analyze alternatives and identify and document the preferred solution.

Part 4: Decision Support Case Using Basic Excel Functionality
Case 10

The book continues with a case that uses basic Excel functionality. (In other words, the case does not require Scenario Manager or the Solver.) Excel is used to test students' analytical skills in "what-if" analyses.

Part 5: Integration Cases Using Access and Excel
Cases 11 and 12

These cases integrate Access and Excel. The cases show students how to share data between Access and Excel to solve problems.

Part 6: Advanced Skills Using Excel

This part contains one tutorial that focuses on using advanced techniques in Excel.

Tutorial E: Guidance for Excel Cases

Some cases require the use of Excel techniques that are not discussed in other tutorials or cases in this casebook. For example, techniques for using data tables and pivot tables are explained in Tutorial E rather than in the cases themselves.

Part 7: Presentation Skills

Tutorial F: Giving an Oral Presentation

Each case includes an optional assignment that lets students practice making a presentation to management to summarize the results of their case analysis. This tutorial gives advice for creating oral presentations. It also includes technical information on charting, a technique that is useful in case analyses or as support for presentations. This tutorial will help students to organize their recommendations, to present their solutions both in words and graphics, and to answer questions from the audience. For larger classes, instructors may want to have students work in teams to create and deliver their presentations, which would model the team approach used by many corporations.

To view and access additional cases, instructors should consider using the "Hall of Fame," as described in the *Using the Cases* section below.

INDIVIDUAL CASE DESIGN

The format of the cases uses the following template:

- Each case begins with a *Preview* and an overview of the tasks.
- The next section, *Preparation*, tells students what they need to do or know to complete the case successfully. Again, the tutorials also prepare students for the cases.
- The third section, *Background*, provides the business context that frames the case. The background of each case models situations that require the kinds of thinking and analysis that students will need in the business world.
- The *Assignment* sections are generally organized to help students develop their analyses.
- The last section, *Deliverables*, lists the finished materials that students must hand in: printouts, a memorandum, a presentation, and files. The list is similar to the deliverables that a business manager might demand.

USING THE CASES

We have successfully used cases like these in our undergraduate MIS courses. We usually begin the semester with Access database instruction. We assign the Access database tutorials and then a case to each student. Then, to teach students how to use the Excel decision support system, we do the same thing: we assign a tutorial and then a case.

Some instructors have asked for access to extra cases, especially in the second semester of a school year. For example, they assigned the integration case in the fall, and they need another one for the spring. To meet this need, we have set up an online "Hall of Fame" that features some of our favorite cases from prior editions. These password-protected cases are available to instructors on the Cengage Learning Web site. Go to *www.cengage.com/coursetechnology* and search for this textbook by title, author, or ISBN. Note that the cases are in Microsoft Office 2010 format.

TECHNICAL INFORMATION

This textbook was tested for quality assurance using the Windows 7 operating system, Microsoft Access 2010, and Microsoft Excel 2010.

Data Files and Solution Files

We have created "starter" data files for the Excel cases, so students need not spend time typing in the spreadsheet skeleton. Cases 11 and 12 also require students to load an Access database file. All these files are on the Cengage Learning Web site, which is available both to students and instructors. Instructors should go to *www.cengage.com* and search for this textbook by title, author, or ISBN. Students will find the files at *www.cengagebrain.com*. You are granted a license to copy the data files to any computer or computer network used by people who have purchased this textbook.

Solutions to the material in the text are available to instructors at *login.cengage.com/sso*. Search for this textbook by title, author, or ISBN. The solutions are password protected.

ACKNOWLEDGMENTS

We would like to give many thanks to the team at Cengage Learning, including our Development Editor, Dan Seiter; Senior Product Manager, Kate Hennessy Mason; and our Content Project Manager, Divya Divakaran. As always, we acknowledge our students' diligent work.

PROBLEM-SOLVING CASES IN
MICROSOFT® ACCESS™ AND EXCEL®

PART 1

DATABASE CASES USING ACCESS

DATABASE DESIGN

This tutorial has three sections. The first section briefly reviews basic database terminology. The second section teaches database design. The third section features a database design problem for practice.

REVIEW OF TERMINOLOGY

You will begin by reviewing some basic terms that will be used throughout this textbook. In Access, a **database** is a group of related objects that are saved in one file. An Access **object** can be a table, form, query, or report. You can identify an Access database file by its suffix, .accdb.

A **table** consists of data that is arrayed in rows and columns. A **row** of data is called a **record**. A **column** of data is called a **field**. Thus, a record is a set of related fields. The fields in a table should be related to one another in some way. For example, a company might want to keep its employee data together by creating a database table called Employee. That table would contain data fields about employees, such as their names and addresses. It would not have data fields about the company's customers; that data would go in a Customer table.

A field's values have a **data type** that is declared when the table is defined. Thus, when data is entered into the database, the software knows how to interpret each entry. Data types in Access include the following:

- Text for words
- Integer for whole numbers
- Double for numbers that have a decimal value
- Currency for numbers that represent dollars and cents
- Yes/No for variables that have only two values (such as 1/0, on/off, yes/no, and true/false)
- Date/Time for variables that are dates or times

Each database table should have a **primary key** field—a field in which each record has a *unique* value. For example, in an Employee table, a field called Employee Identification Number (EIN) could serve as a primary key. (This assumes that each employee is given a number when hired, and that these numbers are not reused later.) Sometimes a table does not have a single field whose values are all different. In that case, two or more fields are combined into a **compound primary key**. The combination of the fields' values is unique.

Database tables should be logically related to one another. For example, suppose a company has an Employee table with fields for EIN, Name, Address, and Telephone Number. For payroll purposes, the company has an Hours Worked table with a field that summarizes Labor Hours for individual employees. The relationship between the Employee table and Hours Worked table needs to be established in the database so you can determine the number of hours worked by any employee. To create this relationship, you include the primary key field from the Employee table (EIN) as a field in the Hours Worked table. In the Hours Worked table, the EIN field is then called a **foreign key** because it's from a "foreign" table.

In Access, data can be entered directly into a table or it can be entered into a form, which then inserts the data into a table. A **form** is a database object that is created from an existing table to make the process of entering data more user-friendly.

A **query** is the database equivalent of a question that is posed about data in a table (or tables). For example, suppose a manager wants to know the names of employees who have worked for the company for more than five years. A query could be designed to search the Employee table for the information. The query would be run, and its output would answer the question.

Queries can be designed to search multiple tables at a time. For this to work, the tables must be connected by a **join** operation, which links tables on the values in a field that they have in common. The common field acts as a "hinge" for the joined tables; when the query is run, the query generator treats the joined tables as one large table.

In Access, queries that answer a question are called *select queries* because they select relevant data from the database records. Queries also can be designed to change data in records, add a record to the end of a table, or delete entire records from a table. These queries are called **update**, **append**, and **delete** queries, respectively.

Access has a **report** generator that can be used to format a table's data or a query's output.

DATABASE DESIGN

Designing a database involves determining which tables belong in the database and then creating the fields that belong in each table. This section begins with an introduction to key database design concepts, then discusses design rules you should use when building a database. First, the following key concepts are defined:

- Entities
- Relationships
- Attributes

Database Design Concepts

Computer scientists have highly formalized ways of documenting a database's logic. Learning their notations and mechanics can be time-consuming and difficult. In fact, doing so usually takes a good portion of a systems analysis and design course. This tutorial will teach you database design by emphasizing practical business knowledge; the approach should enable you to design serviceable databases quickly. Your instructor may add more formal techniques.

A database models the logic of an organization's operation, so your first task is to understand the operation. You can talk to managers and workers, make your own observations, and look at business documents such as sales records. Your goal is to identify the business's "entities" (sometimes called *objects*). An **entity** is a thing or event that the database will contain. Every entity has characteristics, called **attributes**, and one or more **relationships** to other entities. Take a closer look.

Entities

As previously mentioned, an entity is a tangible thing or an event. The reason for identifying entities is that *an entity eventually becomes a table in the database*. Entities that are things are easy to identify. For example, consider a video store. The database for the video store would probably need to contain the names of DVDs and the names of customers who rent them, so you would have one entity named Video and another named Customer.

In contrast, entities that are events can be more difficult to identify, probably because they are more conceptual. However, events are real, and they are important. In the video store example, one event would be Video Rental and another event would be Hours Worked by employees.

In general, your analysis of an organization's operations is made easier when you realize that organizations usually have physical entities such as these:

- Employees
- Customers
- Inventory (products or services)
- Suppliers

Thus, the database for most organizations would have a table for each of these entities. Your analysis also can be made easier by knowing that organizations engage in transactions internally (within the company) and externally (with the outside world). Such transactions are explained in an introductory accounting course, but most people understand them from events that occur in daily life. Consider the following examples:

- Organizations generate revenue from sales or interest earned. Revenue-generating transactions include event entities called Sales and Interest Earned.
- Organizations incur expenses from paying hourly employees and purchasing materials from suppliers. Hours Worked and Purchases are event entities in the databases of most organizations.

Thus, identifying entities is a matter of observing what happens in an organization. Your powers of observation are aided by knowing what entities exist in the databases of most organizations.

Relationships

As an analyst building a database, you should consider the relationship of each entity to the other entities you have identified. For example, a college database might contain entities for Student, Course, and Section to contain data about each. A relationship between Student and Section could be expressed as "Students enroll in sections."

An analyst also must consider the **cardinality** of any relationship. Cardinality can be one-to-one, one-to-many, or many-to-many:

- In a one-to-one relationship, one instance of the first entity is related to just one instance of the second entity.
- In a one-to-many relationship, one instance of the first entity is related to many instances of the second entity, but each instance of the second entity is related to only one instance of the first.
- In a many-to-many relationship, one instance of the first entity is related to many instances of the second entity, and one instance of the second entity is related to many instances of the first.

For a more concrete understanding of cardinality, consider again the college database with the Student, Course, and Section entities. The university catalog shows that a course such as Accounting 101 can have more than one section: 01, 02, 03, 04, and so on. Thus, you can observe the following relationships:

- The relationship between the entities Course and Section is one-to-many. Each course has many sections, but each section is associated with just one course.
- The relationship between Student and Section is many-to-many. Each student can be in more than one section, because each student can take more than one course. Also, each section has more than one student.

Thinking about relationships and their cardinalities may seem tedious to you. However, as you work through the cases in this text, you will see that this type of analysis can be valuable in designing databases. In the case of many-to-many relationships, you should determine the tables a given database needs; in the case of one-to-many relationships, you should decide which fields the tables need to share.

Attributes

An attribute is a characteristic of an entity. You identify attributes of an entity because *attributes become a table's fields*. If an entity can be thought of as a noun, an attribute can be considered an adjective that describes the noun. Continuing with the college database example, consider the Student entity. Students have names, so Last Name would be an attribute of the Student entity and therefore a field in the Student table. First Name would be another attribute, as well as Address, Phone Number, and other descriptive fields.

Sometimes it can be difficult to tell the difference between an attribute and an entity, but one good way is to ask whether more than one attribute is possible for each entity. If more than one instance is possible, but you do not know the number in advance, you are working with an entity. For example, assume that a student could have a maximum of two addresses—one for home and one for college. You could specify attributes Address 1 and Address 2. Next, consider that you might not know the number of student addresses in advance, meaning that all addresses have to be recorded. In that case, you would not know how many fields to set aside in the Student table for addresses. Therefore, you would need a separate Student Addresses table (entity) that would show any number of addresses for a given student.

Database Design Rules

As described previously, your first task in database design is to understand the logic of the business situation. Once you understand this logic, you are ready to build the database. To create a context for learning about database design, look at a hypothetical business operation and its database needs.

Example: The Talent Agency

Suppose you have been asked to build a database for a talent agency that books musical bands into nightclubs. The agent needs a database to keep track of the agency's transactions and to answer day-to-day questions. For example, a club manager often wants to know which bands are available on a certain date at a certain time, or wants to know the agent's fee for a certain band. The agent may want to see a list of all band members and the instrument each person plays, or a list of all bands that have three members.

Suppose that you have talked to the agent and have observed the agency's business operation. You conclude that your database needs to reflect the following facts:

1. A booking is an event in which a certain band plays in a particular club on a particular date, starting and ending at certain times, and performing for a specific fee. A band can play more than once a day. The Heartbreakers, for example, could play at the East End Cafe in the afternoon and then at the West End Cafe on the same night. For each booking, the club pays the talent agent. The agent keeps a five percent fee and then gives the remainder of the payment to the band.

2. Each band has at least two members and an unlimited maximum number of members. The agent notes a telephone number of just one band member, which is used as the band's contact number. No two bands have the same name or telephone number.

3. Band member names are not unique. For example, two bands could each have a member named Sally Smith.

4. The agent keeps track of just one instrument that each band member plays. For the purpose of this database, "vocals" are considered an instrument.

5. Each band has a desired fee. For example, the Lightmetal band might want $700 per booking, and would expect the agent to try to get at least that amount.

6. Each nightclub has a name, an address, and a contact person. The contact person has a telephone number that the agent uses to call the club. No two clubs have the same name, contact person, or telephone number. Each club has a target fee. The contact person will try to get the agent to accept that fee for a band's appearance.

7. Some clubs feed the band members for free; others do not.

Before continuing with this tutorial, you might try to design the agency's database on your own. Ask yourself: What are the entities? Recall that business databases usually have Customer, Employee, and Inventory entities, as well as an entity for the event that generates revenue transactions. Each entity becomes a table in the database. What are the relationships among the entities? For each entity, what are its attributes? For each table, what is the primary key?

Six Database Design Rules

Assume that you have gathered information about the business situation in the talent agency example. Now you want to identify the tables required for the database and the fields needed in each table. Observe the following six rules:

Rule 1: You do not need a table for the business. The database represents the entire business. Thus, in the example, Agent and Agency are not entities.

Rule 2: Identify the entities in the business description. Look for typical things and events that will become tables in the database. In the talent agency example, you should be able to observe the following entities:

- *Things*: The product (inventory for sale) is Band. The customer is Club.
- *Events*: The revenue-generating transaction is Bookings.

You might ask yourself: Is there an Employee entity? Isn't Instrument an entity? Those issues will be discussed as the rules are explained.

Rule 3: Look for relationships among the entities. Look for one-to-many relationships between entities. The relationship between those entities must be established in the tables, using a foreign key. For details, see the following discussion in Rule 4 about the relationship between Band and Band Member.

Look for many-to-many relationships between entities. Each of these relationships requires a third entity that associates the two entities in the relationship. Recall the many-to-many relationship from the college database scenario that involved Student and Section entities. To display the enrollment of specific students in specific sections, a third table would be required. The mechanics of creating such a table are described in Rule 4 during the discussion of the relationship between Band and Club.

Rule 4: Look for attributes of each entity and designate a primary key. As previously mentioned, you should think of the entities in your database as nouns. You should then create a list of adjectives that describe those nouns. These adjectives are the attributes that will become the table's fields. After you have identified fields for each table, you should check to see whether a field has unique values. If such a field exists, designate it as the primary key field; otherwise, designate a compound primary key.

In the talent agency example, the attributes, or fields, of the Band entity are Band Name, Band Phone Number, and Desired Fee, as shown in Figure A-1. Assume that no two bands have the same name, so the primary key field can be Band Name. The data type of each field is shown.

BAND	
Field Name	**Data Type**
Band Name (primary key)	Text
Band Phone Number	Text
Desired Fee	Currency

FIGURE A-1 The Band table and its fields

Two Band records are shown in Figure A-2.

Band Name (primary key)	Band Phone Number	Desired Fee
Heartbreakers	981 831 1765	$800
Lightmetal	981 831 2000	$700

FIGURE A-2 Records in the Band table

If two bands might have the same name, Band Name would not be a good primary key, so a different unique identifier would be needed. Such situations are common. Most businesses have many types of inventory, and duplicate names are possible. The typical solution is to assign a number to each product to use as the primary key field. For example, a college could have more than one faculty member with the same name, so each faculty member would be assigned an employee identification number. Similarly, banks assign a personal identification number (PIN) for each depositor. Each automobile produced by a car manufacturer gets a unique Vehicle Identification Number (VIN). Most businesses assign a number to each sale, called an invoice number. (The next time you go to a grocery store, note the number on your receipt. It will be different from the number on the next customer's receipt.)

At this point, you might be wondering why Band Member would not be an attribute of Band. The answer is that, although you must record each band member, you do not know in advance how many members are in each band. Therefore, you do not know how many fields to allocate to the Band table for members. (Another way to think about band members is that they are the agency's employees, in effect. Databases for organizations usually have an Employee entity.) You should create a Band Member table with the attributes Member ID Number, Member Name, Band Name, Instrument, and Phone. A Member ID Number field is needed because member names may not be unique. The table and its fields are shown in Figure A-3.

BAND MEMBER	
Field Name	**Data Type**
Member ID Number (primary key)	Text
Member Name	Text
Band Name (foreign key)	Text
Instrument	Text
Phone	Text

FIGURE A-3 The Band Member table and its fields

Note in Figure A-3 that the phone number is classified as a Text data type because the field values will not be used in an arithmetic computation. The benefit is that Text data type values take up fewer bytes than Numerical or Currency data type values; therefore, the file uses less storage space. You should also use the Text data type for number values such as zip codes.

Five records in the Band Member table are shown in Figure A-4.

Member ID Number	Member Name	Band Name	Instrument	Phone
0001	Pete Goff	Heartbreakers	Guitar	981 444 1111
0002	Joe Goff	Heartbreakers	Vocals	981 444 1234
0003	Sue Smith	Heartbreakers	Keyboard	981 555 1199
0004	Joe Jackson	Lightmetal	Sax	981 888 1654
0005	Sue Hoopes	Lightmetal	Piano	981 888 1765

FIGURE A-4   Records in the Band Member table

You can include Instrument as a field in the Band Member table because the agent records only one instrument for each band member. Thus, you can use the instrument as a way to describe a band member, much like the phone number is part of the description. Phone could not be the primary key because two members might share a telephone and because members might change their numbers, making database administration more difficult.

You might ask why Band Name is included in the Band Member table. The common-sense reason is that you did not include the Member Name in the Band table. You must relate bands and members somewhere, and the Band Member table is the place to do it.

To think about this relationship in another way, consider the cardinality of the relationship between Band and Band Member. It is a one-to-many relationship: one band has many members, but each member in the database plays in just one band. You establish such a relationship in the database by using the primary key field of one table as a foreign key in the other table. In Band Member, the foreign key Band Name is used to establish the relationship between the member and his or her band.

The attributes of the Club entity are Club Name, Address, Contact Name, Club Phone Number, Preferred Fee, and Feed Band?. The Club table can define the Club entity, as shown in Figure A-5.

CLUB	
Field Name	**Data Type**
Club Name (primary key)	Text
Address	Text
Contact Name	Text
Club Phone Number	Text
Preferred Fee	Currency
Feed Band?	Yes/No

FIGURE A-5   The Club table and its fields

Two records in the Club table are shown in Figure A-6.

Club Name (primary key)	Address	Contact Name	Club Phone Number	Preferred Fee	Feed Band?
East End	1 Duce St.	Al Pots	981 444 8877	$600	Yes
West End	99 Duce St.	Val Dots	981 555 0011	$650	No

FIGURE A-6 Records in the Club table

You might wonder why Bands Booked into Club (or a similar name) is not an attribute of the Club table. There are two reasons. First, you do not know in advance how many bookings a club will have, so the value cannot be an attribute. Second, Bookings is the agency's revenue-generating transaction, an event entity, and you need a table for that business transaction. Consider the booking transaction next.

You know that the talent agent books a certain band into a certain club for a specific fee on a certain date, starting and ending at a specific time. From that information, you can see that the attributes of the Bookings entity are Band Name, Club Name, Date, Start Time, End Time, and Fee. The Bookings table and its fields are shown in Figure A-7.

BOOKINGS	
Field Name	Data Type
Band Name	Text
Club Name	Text
Date	Date/Time
Start Time	Date/Time
End Time	Date/Time
Fee	Currency

FIGURE A-7 The Bookings table and its fields—and no designation of a primary key

Some records in the Bookings table are shown in Figure A-8.

Band Name	Club Name	Date	Start Time	End Time	Fee
Heartbreakers	East End	11/21/12	21:30	23:30	$800
Heartbreakers	East End	11/22/12	21:00	23:30	$750
Heartbreakers	West End	11/28/12	19:00	21:00	$500
Lightmetal	East End	11/21/12	18:00	20:00	$700
Lightmetal	West End	11/22/12	19:00	21:00	$750

FIGURE A-8 Records in the Bookings table

Note that no single field is guaranteed to have unique values, because each band is likely to be booked many times and each club might be used many times. Furthermore, each date and time can appear more than once. Thus, no one field can be the primary key.

If a table does not have a single primary key field, you can make a compound primary key whose field values will be unique when taken together. Because a band can be in only one place at a time, one possible solution is to create a compound key from the Band Name, Date, and Start Time fields. An alternative solution is to create a compound primary key from the Club Name, Date, and Start Time fields.

If you don't want a compound key, you could create a field called Booking Number. Each booking would then have its own unique number, similar to an invoice number.

You can also think about this event entity in a different way. Over time, a band plays in many clubs, and each club hires many bands. Thus, Band and Club have a many-to-many relationship, which signals the need for a table between the two entities. A Bookings table would associate the Band and Club tables. You implement an associative table by including the primary keys from the two tables that are associated. In this case, the primary keys from the Band and Club tables are included as foreign keys in the Bookings table.

Rule 5: *Avoid data redundancy.* You should not include extra (redundant) fields in a table. Redundant fields take up extra disk space and lead to data entry errors because the same value must be entered in multiple tables, increasing the chance of a keystroke error. In large databases, keeping track of multiple instances of the same data is nearly impossible, so contradictory data entries become a problem.

Consider this example: Why wouldn't Club Phone Number be included in the Bookings table as a field? After all, the agent might have to call about a last-minute booking change and could quickly look up the number in the Bookings table. Assume that the Bookings table includes Booking Number as the primary key and Club Phone Number as a field. Figure A-9 shows the Bookings table with the additional field.

BOOKINGS	
Field Name	**Data Type**
Booking Number (primary key)	Text
Band Name	Text
Club Name	Text
Club Phone Number	Text
Date	Date/Time
Start Time	Date/Time
End Time	Date/Time
Fee	Currency

FIGURE A-9   The Bookings table with an unnecessary field—Club Phone Number

The fields Date, Start Time, End Time, and Fee logically depend on the Booking Number primary key—they help define the booking. Band Name and Club Name are foreign keys and are needed to establish the relationship between the Band, Club, and Bookings tables. But what about Club Phone Number? It is not defined by the Booking Number. It is defined by Club Name—*in other words, it is a function of the club, not of the booking.* Thus, the Club Phone Number field does not belong in the Bookings table. It is already in the Club table.

Perhaps you can see the practical data-entry problem of including Club Phone Number in Bookings. Suppose a club changed its contact phone number. The agent could easily change the number one time, in the Club table. However, the agent would need to remember which other tables contained the field and change the values there too. In a small database, this task might not be difficult, but in larger databases, having redundant fields in many tables makes such maintenance difficult, which means that redundant data is often incorrect.

You might object by saying, "What about all of those foreign keys? Aren't they redundant?" In a sense, they are. But they are needed to establish the relationship between one entity and another, as discussed previously.

Rule 6: *Do not include a field if it can be calculated from other fields.* A **calculated field** is made using the query generator. Thus, the agent's fee is not included in the Bookings table because it can be calculated by query (here, five percent multiplied by the booking fee).

PRACTICE DATABASE DESIGN PROBLEM

Imagine that your town library wants to keep track of its business in a database, and that you have been called in to build the database. You talk to the town librarian, review the old paper-based records, and watch people use the library for a few days. You learn the following about the library:

1. Any resident of the town can get a library card simply by asking for one. The library considers each cardholder a member of the library.

2. The librarian wants to be able to contact members by telephone and by mail. She calls members when books are overdue or when requested materials become available. She likes to mail a thank-you note to each patron on his or her anniversary of becoming a member of the library. Without a database, contacting members efficiently can be difficult; for example, multiple members can have the same name. Also, a parent and a child might have the same first and last name, live at the same address, and share a phone.

3. The librarian tries to keep track of each member's reading interests. When new books come in, the librarian alerts members whose interests match those books. For example, long-time member Sue Doaks is interested in reading Western novels, growing orchids, and baking bread. There must be some way to match her interests with available books. One complication is that, although the librarian wants to track all of a member's reading interests, she wants to classify each book as being in just one category of interest. For example, the classic gardening book *Orchids of France* would be classified as a book about orchids or a book about France, but not both.

4. The library stocks thousands of books. Each book has a title and any number of authors. Also, more than one book in the library might have the same title. Similarly, multiple authors might have the same name.

5. A writer could be the author of more than one book.

6. A book will be checked out repeatedly as time goes on. For example, *Orchids of France* could be checked out by one member in March, by another member in July, and by another member in September.

7. The library must be able to identify whether a book is checked out.

8. A member can check out any number of books in one visit. Also, a member might visit the library more than once a day to check out books.

9. All books that are checked out are due back in two weeks, with no exceptions. The librarian would like to have an automated way of generating an overdue book list each day so she can telephone offending members.

10. The library has a number of employees. Each employee has a job title. The librarian is paid a salary, but other employees are paid by the hour. Employees clock in and out each day. Assume that all employees work only one shift per day and that all are paid weekly. Pay is deposited directly into an employee's checking account—no checks are hand-delivered. The database needs to include the librarian and all other employees.

Design the library's database, following the rules set forth in this tutorial. Your instructor will specify the format of your work. Here are a few hints in the form of questions:

- A book can have more than one author. An author can write more than one book. How would you describe the relationship between books and authors?

- The library lends books for free, of course. If you were to think of checking out a book as a sales transaction for zero revenue, how would you handle the library's revenue-generating event?

- A member can borrow any number of books at one checkout. A book can be checked out more than once. How would you describe the relationship between checkouts and books?

TUTORIAL

MICROSOFT ACCESS

Microsoft Access is a relational database package that runs on the Microsoft Windows operating system. There are many different versions of Access; this tutorial was prepared using Access 2010.

Before using this tutorial, you should know the fundamentals of Access and know how to use Windows. This tutorial explains advanced Access skills you will need to complete database case studies. The tutorial concludes with a discussion of common Access problems and how to solve them.

To prevent losing your work, always observe proper file-saving and closing procedures. To exit Access, click the File tab and select Close Database, then click the File tab and select Exit. You can also simply select the Exit option to return to Windows. Always end your work with these steps. If you remove your USB key or other portable storage device when database forms and tables are shown on the screen, you will lose your work.

To begin this tutorial, you will create a new database called Employee.

AT THE KEYBOARD

Open a new database. Click the File tab, select New from the menu, and then click Blank database from the Available Templates list. Name the database Employee. Click the file folder next to the filename to browse for the folder where you want to save the file. Otherwise, your file will be saved automatically in the Documents folder. Click the Create button.

Your opening screen should resemble the screen shown in Figure B-1.

FIGURE B-1   Entering data in Datasheet view

When you create a table, Access opens it in Datasheet view by default. Because you will use Design view to build your tables, close the new table by clicking the *X* in the upper-right corner of the table window that corresponds to Close Table I. You are now on the Home tab in the Database window of Access, as shown in Figure B-2. From this screen, you can create or change objects.

FIGURE B-2   The Database window Home tab in Access

CREATING TABLES

Your database will contain data about employees, their wage rates, and the hours they worked.

Defining Tables

In the Database window, build three new tables using the following instructions.

AT THE KEYBOARD

Defining the Employee Table

This table contains permanent data about employees. To create the table, click the Create tab and then click Table Design in the Tables group. The table's fields are Last Name, First Name, Employee ID, Street Address, City, State, Zip, Date Hired, and US Citizen. The Employee ID field is the primary key field. Change the lengths of text fields from the default 255 spaces to more appropriate lengths; for example, the Last Name field might be 30 spaces, and the Zip field might be 10 spaces. Your completed definition should resemble the one shown in Figure B-3.

Field Name	Data Type	Description
Last Name	Text	
First Name	Text	
⚷ Employee ID	Text	
Street Address	Text	
City	Text	
State	Text	
Zip	Text	
Date Hired	Date/Time	
US Citizen	Yes/No	

FIGURE B-3 Fields in the Employee table

When you finish, click the File tab, select Save Object As, and then enter a name for the table. In this example, the table is named Employee. Make sure to specify the name of the *table*, not the database itself. (In this example, it is a coincidence that the Employee table has the same name as its database file.) Close the table by clicking the Close button (X) that corresponds to the Employee table.

Defining the Wage Data Table

This table contains permanent data about employees and their wage rates. The table's fields are Employee ID, Wage Rate, and Salaried. The Employee ID field is the primary key field. Use the data types shown in Figure B-4. Your definition should resemble the one shown in Figure B-4.

Field Name	Data Type	Description
⚷ Employee ID	Text	
Wage Rate	Currency	
Salaried	Yes/No	

FIGURE B-4 Fields in the Wage Data table

Click the File tab and then select Save Object As to save the table definition. Name the table Wage Data.

Defining the Hours Worked Table

The purpose of this table is to record the number of hours that employees work each week during the year. The table's three fields are Employee ID (which has a text data type), Week # (number–long integer), and Hours (number–double). The Employee ID and Week # are the compound keys.

In the following example, the employee with ID number 08965 worked 40 hours in Week 1 of the year and 52 hours in Week 2.

Employee ID	Week#	Hours
08965	1	40
08965	2	52

Note that no single field can be the primary key field because 08965 is an entry for each week. In other words, if this employee works each week of the year, 52 records will have the same Employee ID value at the end of the year. Thus, Employee ID values will not distinguish records. No other single field can distinguish these records either, because other employees will have worked during the same week number and some employees will have worked the same number of hours. For example, 40 hours—which corresponds to a full-time workweek—would be a common entry for many weeks.

All of this presents a problem because a table must have a primary key field in Access. The solution is to use a compound primary key; that is, use values from more than one field to create a combined field that will distinguish records. The best compound key to use for the current example consists of the Employee ID field and the Week # field, because as each person works each week, the week number changes. In other words, there is only *one* combination of Employee ID 08965 and Week # 1. Because those values *can occur in only one record*, the combination distinguishes that record from all others.

The first step of setting a compound key is to highlight the fields in the key. Those fields must appear one after the other in the table definition screen. (Plan ahead for that format.) As an alternative, you can highlight one field, hold down the Control key, and highlight the next field.

AT THE KEYBOARD

In the Hours Worked table, click the first field's left prefix area (known as the row selector), hold down the mouse button, and drag down to highlight the names of all fields in the compound primary key. Your screen should resemble the one shown in Figure B-5.

Field Name	Data Type	Description
Employee ID	Text	
Week #	Number	
Hours	Number	

FIGURE B-5   Selecting fields for the compound primary key for the Hours Worked table

Now click the Key icon. Your screen should resemble the one shown in Figure B-6.

Field Name	Data Type	Description
Employee ID	Text	
Week #	Number	
Hours	Number	

FIGURE B-6   The compound primary key for the Hours Worked table

You have created the compound primary key and finished defining the table. Click the File tab and then select Save Object As to save the table as Hours Worked.

Adding Records to a Table

At this point, you have set up the skeletons of three tables. The tables have no data records yet. If you printed the tables now, you would only see column headings (the field names). The most direct way to enter data into a table is to double-click the table's name in the navigation pane at the left side of the screen and then type the data directly into the cells.

NOTE

To display and open the database objects, Access 2010 uses a navigation pane, which is on the left side of the Access window.

AT THE KEYBOARD

On the Home tab of the Database window, double-click the Employee table. Your data entry screen should resemble the one shown in Figure B-7.

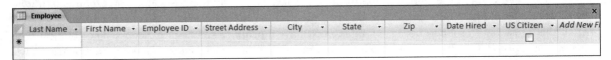

Last Name ▾	First Name ▾	Employee ID ▾	Street Address ▾	City ▾	State ▾	Zip ▾	Date Hired ▾	US Citizen ▾	Add New F
*								☐	

FIGURE B-7   The data entry screen for the Employee table

The Employee table has many fields, some of which may be off the screen to the right. Scroll to see obscured fields. (Scrolling happens automatically as you enter data.) Figure B-7 shows all of the fields on the screen.

Enter your data one field value at a time. Note that the first row is empty when you begin. Each time you finish entering a value, press Enter to move the cursor to the next cell. After you enter data in the last cell in a row, the cursor moves to the first cell of the next row *and* Access automatically saves the record. Thus, you do not need to click the File tab and then select Save Object As after entering data into a table.

When entering data in your table, you should enter dates in the following format: 6/15/10. Access automatically expands the entry to the proper format in output.

Also note that Yes/No variables are clicked (checked) for Yes; otherwise, the box is left blank for No. You can change the box from Yes to No by clicking it.

Enter the data shown in Figure B-8 into the Employee table. If you make errors in data entry, click the cell, backspace over the error, and type the correction.

Last Name ▾	First Name ▾	Employee ID ▾	Street Address ▾	City ▾	State ▾	Zip ▾	Date Hired ▾	US Citizen ▾
Howard	Jane	11411	28 Sally Dr	Glasgow	DE	19702	8/1/2011	☑
Smith	John	12345	30 Elm St	Newark	DE	19711	6/1/1996	☑
Smith	Albert	14890	44 Duce St	Odessa	DE	19722	7/15/1987	☑
Jones	Sue	22282	18 Spruce St	Newark	DE	19716	7/15/2004	☐
Ruth	Billy	71460	1 Tater Dr	Baltimore	MD	20111	8/15/1999	☐
Add	Your	Data	Here	Elkton	MD	21921		☑
*								☐

FIGURE B-8   Data for the Employee table

Note that the sixth record is *your* data record. Assume that you live in Elkton, Maryland, were hired on today's date (enter the date), and are a U.S. citizen. Make up a fictitious Employee ID number. For purposes of this tutorial, the sixth record has been created using the name of one of this text's authors and the employee ID 09911.

After adding records to the Employee table, open the Wage Data table and enter the data shown in Figure B-9.

Employee ID ▾	Wage Rate ▾	Salaried ▾
11411	$10.00	☐
12345		☑
14890	$12.00	☐
22282		☑
71460		☑
Your Employee ID	$8.00	☐
*		☐

FIGURE B-9   Data for the Wage Data table

In this table, you are again asked to create a new entry. For this record, enter your own employee ID. Also assume that you earn $8 an hour and are not salaried. Note that when an employee's Salaried box is not checked (in other words, Salaried = No), the implication is that the employee is paid by the hour. Because salaried employees are not paid by the hour, their hourly rate is 0.00.

When you finish creating the Wage Data table, open the Hours Worked table and enter the data shown in Figure B-10.

▦ Hours Worked		
Employee ID ▾	Week # ▾	Hours ▾
11411	1	40
11411	2	50
12345	1	40
12345	2	40
14890	1	38
14890	2	40
22282	1	40
22282	2	40
71460	1	40
71460	2	40
Your Employee ID	1	60
Your Employee ID	2	55
*		

FIGURE B-10   Data for the Hours Worked table

Notice that salaried employees are always given 40 hours. Nonsalaried employees (including you) might work any number of hours. For your record, enter your fictitious employee ID, 60 hours worked for Week 1, and 55 hours worked for Week 2.

CREATING QUERIES

Because you know how to create basic queries, this section explains the advanced queries you will create in the cases in this book.

Using Calculated Fields in Queries

A **calculated field** is an output field made up of *other* field values. A calculated field is *not* a field in a table; it is created in the query generator. The calculated field does not become part of the table—it is just part of the query output. The best way to understand this process is to work through an example.

AT THE KEYBOARD

Suppose you want to see the employee IDs and wage rates of hourly workers, and the new wage rates if all employees were given a 10 percent raise. To view that information, show the employee ID, the current wage rate, and the higher rate, which should be titled New Rate in the output. Figure B-11 shows how to set up the query.

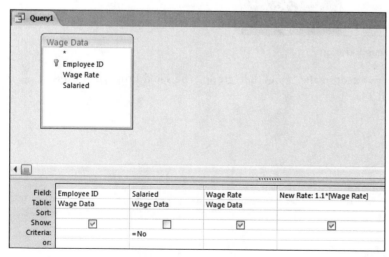

FIGURE B-11   Query setup for the calculated field

To set up this query, you need to select hourly workers by using the Salaried field with Criteria = No. Note in Figure B-11 that the Show box for the field is not checked, so the Salaried field values will not appear in the query output.

Note the expression for the calculated field, which you can see in the far-right field cell:

New Rate: 1.1 * [Wage Rate]

The term *New Rate:* merely specifies the desired output heading. (Don't forget the colon.) The rest of the expression, 1.1 * [Wage Rate], multiplies the old wage rate by 110 percent, which results in the 10 percent raise.

In the expression, the field name Wage Rate must be enclosed in square brackets. Remember this rule: *Any time an Access expression refers to a field name, the expression must be enclosed in square brackets.*

If you run this query, your output should resemble that in Figure B-12.

Query1		
Employee ID ▾	Wage Rate ▾	New Rate ▾
11411	$10.00	11
14890	$12.00	13.2
09911	$8.00	8.8
*		

FIGURE B-12   Output for a query with calculated field

Notice that the calculated field output is not shown in Currency format, but as a Double—a number with digits after the decimal point. To convert the output to Currency format, select the output column by clicking the line above the calculated field expression. The column darkens to indicate its selection. Your data entry screen should resemble the one shown in Figure B-13.

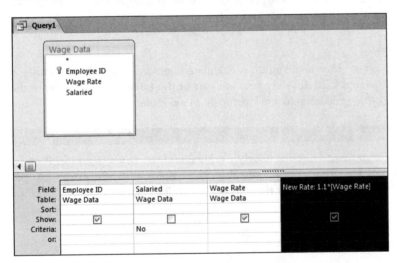

FIGURE B-13   Activating a calculated field in query design

Then, on the Design tab, click Property Sheet in the Show/Hide group. The Field Properties window appears, as shown on the right in Figure B-14.

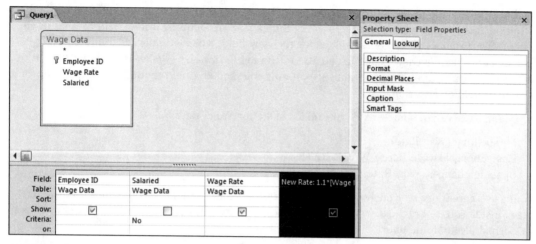

FIGURE B-14   Field properties of a calculated field

Click Format and choose Currency, as shown in Figure B-15. Then click the *X* in the upper-right corner of the window to close it.

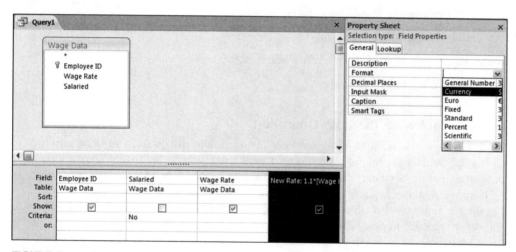

FIGURE B-15   Currency format of a calculated field

When you run the query, the output should resemble that in Figure B-16.

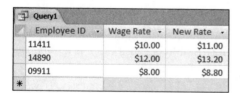

FIGURE B-16   Query output with formatted calculated field

Next, you examine how to avoid errors when making calculated fields.

Avoiding Errors when Making Calculated Fields

Follow these guidelines to avoid making errors in calculated fields:

- Do not enter the expression in the *Criteria* cell as if the field definition were a filter. You are making a field, so enter the expression in the *Field* cell.

- Spell, capitalize, and space a field's name *exactly* as you did in the table definition. If the table definition differs from what you type, Access thinks you are defining a new field by that name. Access then prompts you to enter values for the new field, which it calls a Parameter Query field. This problem is easy to debug because of the tag *Parameter Query*. If Access asks you to enter values for a parameter, you almost certainly misspelled a field name in an expression in a calculated field or criterion.

For example, here are some errors you might make for Wage Rate:

> Misspelling: (Wag Rate)
> Case change: (wage Rate / WAGE RATE)
> Spacing change: (WageRate / Wage Rate)

- Do not use parentheses or curly braces instead of the square brackets. Also, do not put parentheses inside square brackets. You *can*, however, use parentheses outside the square brackets in the normal algebraic manner.

For example, suppose that you want to multiply Hours by Wage Rate to get a field called Wages Owed. This is the correct expression:

> Wages Owed: [Wage Rate] * [Hours]

The following expression also would be correct:

> Wages Owed: ([Wage Rate] * [Hours])

But it would *not* be correct to omit the inside brackets, which is a common error:

> Wages Owed: [Wage Rate * Hours]

"Relating" Two or More Tables by the Join Operation

Often, the data you need for a query is in more than one table. To complete the query, you must **join** the tables by linking the common fields. One rule of thumb is that joins are made on fields that have common *values*, and those fields often can be key fields. The names of the join fields are irrelevant; also, the names of the tables or fields to be joined may be the same, but it is not required for an effective join.

Make a join by bringing in (adding) the tables needed. Next, decide which fields you will join. Then click one field name and hold down the left mouse button while you drag the cursor over to the other field's name in its window. Release the button. Access inserts a line to signify the join. (If a relationship between two tables has been formed elsewhere, Access inserts the line automatically, and you do not have to perform the click-and-drag operation. Access often inserts join lines without the user forming relationships.)

You can join more than two tables. The common fields *need not* be the same in all tables; that is, you can daisy-chain them together.

A common join error is to add a table to the query and then fail to link it to another table. In that case, you will have a table floating in the top part of the QBE (query by example) screen. When you run the query, your output will show the same records over and over. The error is unmistakable because there is *so much* redundant output. The two rules are to add only the tables you need and to link all tables.

Next, you will work through an example of a query that needs a join.

AT THE KEYBOARD

Suppose you want to see the last names, employee IDs, wage rates, salary status, and citizenship only for U.S. citizens and hourly workers. Because the data is spread across two tables, Employee and Wage Data, you should add both tables and pull down the five fields you need. Then you should add the Criteria expressions. Set up your work to resemble that in Figure B-17. Make sure the tables are joined on the common field, Employee ID.

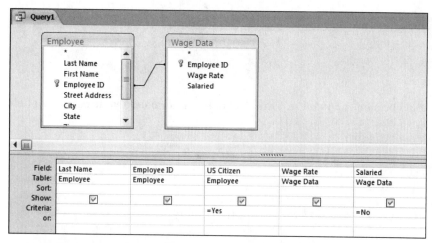

FIGURE B-17   A query based on two joined tables

You should quickly review the criteria you will need to set up this join: If you want data for employees who are U.S. citizens *and* who are hourly workers, the Criteria expressions go in the *same* Criteria row. If you want data for employees who are U.S. citizens *or* who are hourly workers, one of the expressions goes in the second Criteria row (the one with the or: notation).

Now run the query. The output should resemble that in Figure B-18, with the exception of the name "Brady."

Last Name	Employee ID	US Citizen	Wage Rate	Salaried
Howard	11411	☑	$10.00	☐
Smith	14890	☑	$12.00	☐
Brady	09911	☑	$8.00	☐
*		☐		☐

FIGURE B-18   Output of a query based on two joined tables

You do not need to print or save the query output, so return to Design view and close the query. Another practice query follows.

AT THE KEYBOARD

Suppose you want to see the wages owed to hourly employees for Week 2. You should show the last name, the employee ID, the salaried status, the week #, and the wages owed. Wages will have to be a calculated field ([Wage Rate] * [Hours]). The criteria are No for Salaried and 2 for the Week #. (This means that another "And" query is required.) Your query should be set up like the one in Figure B-19.

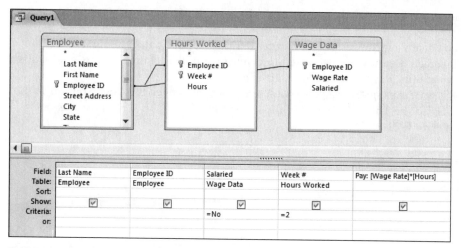

FIGURE B-19   Query setup for wages owed to hourly employees for Week 2

NOTE

In the query in Figure B-19, the calculated field column was widened so you could see the whole expression. To widen a column, click the column boundary line and drag to the right.

Run the query. The output should be similar to that in Figure B-20, if you formatted your calculated field to Currency.

Last Name	Employee ID	Salaried	Week #	Pay
Howard	11411	☐	2	$500.00
Smith	14890	☐	2	$480.00
Brady	09911	☐	2	$440.00
*		☐		

FIGURE B-20 Query output for wages owed to hourly employees for Week 2

Notice that it was not necessary to pull down the Wage Rate and Hours fields to make the query work. You do not need to save or print the query output, so return to Design view and close the query.

Summarizing Data from Multiple Records (Totals Queries)

You may want data that summarizes values from a field for several records (or possibly all records) in a table. For example, you might want to know the average hours that all employees worked in a week or the total (sum) of all of the hours worked. Furthermore, you might want data grouped or stratified in some way. For example, you might want to know the average hours worked, grouped by all U.S. citizens versus all non-U.S. citizens. Access calls such a query a **Totals query**. These queries include the following operations:

Sum	The total of a given field's values
Count	A count of the number of instances in a field—that is, the number of records. In the current example, you would count the number of employee IDs to get the number of employees.
Average	The average of a given field's values
Min	The minimum of a given field's values
Var	The variance of a given field's values
StDev	The standard deviation of a given field's values
Where	The field has criteria for the query output

AT THE KEYBOARD

Suppose you want to know how many employees are represented in the example database. First, bring the Employee table into the QBE screen. Because you will need to count the number of employee IDs, which is a Totals query operation, you must bring down the Employee ID field.

To tell Access that you want a Totals query, click the Design tab and then click the Totals button in the Show/Hide group. A new row called the Total row opens in the lower part of the QBE screen. At this point, the screen resembles that in Figure B-21.

FIGURE B-21   Totals query setup

Note that the Total cell contains the words *Group By*. Until you specify a statistical operation, Access assumes that a field will be used for grouping (stratifying) data.

To count the number of employee IDs, click next to Group By to display an arrow. Click the arrow to reveal a drop-down menu, as shown in Figure B-22.

FIGURE B-22   Choices for statistical operation in a Totals query

Select the Count operator. (You might need to scroll down the menu to see the operator you want.) Your screen should resemble the one shown in Figure B-23.

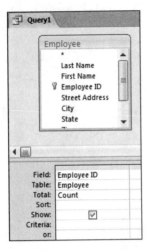

FIGURE B-23   Count in a Totals query

Run the query. Your output should resemble that in Figure B-24.

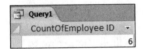

FIGURE B-24   Output of Count in a Totals query

Notice that Access created a pseudo-heading, "CountOfEmployee ID," by splicing together the statistical operation (Count), the word Of, and the name of the field (Employee ID). If you wanted a phrase such as "Count of Employees" as a heading, you would go to Design view and change the query to resemble the one shown in Figure B-25.

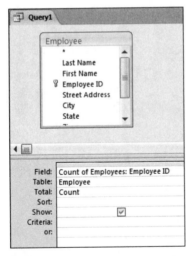

FIGURE B-25   Heading change in a Totals query

When you run the query, the output should resemble that in Figure B-26.

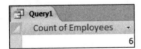

FIGURE B-26   Output of heading change in a Totals query

You do not need to print or save the query output, so return to Design view and close the query.

AT THE KEYBOARD

As another example of a Totals query, suppose you want to know the average wage rate of employees, grouped by whether the employees are salaried. Figure B-27 shows how to set up your query.

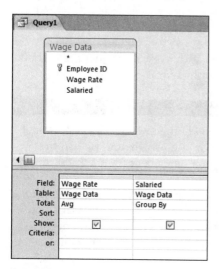

FIGURE B-27   Query setup for average wage rate of employees

When you run the query, your output should resemble that in Figure B-28.

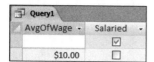

FIGURE B-28   Output of query for average wage rate of employees

Recall the convention that salaried workers are assigned zero dollars an hour. Suppose you want to eliminate the output line for zero dollars an hour because only hourly-rate workers matter for the query. The query setup is shown in Figure B-29.

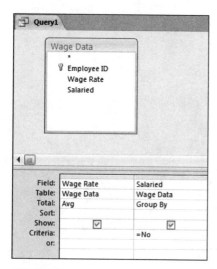

FIGURE B-29   Query setup for nonsalaried workers only

When you run the query, you will get output for nonsalaried employees only, as shown in Figure B-30.

AvgOfWage ▾	Salaried ▾
$10.00	☐

FIGURE B-30   Query output for nonsalaried workers only

Thus, it is possible to use Criteria in a Totals query, just as you would with a "regular" query. You do not need to print or save the query output, so return to Design view and close the query.

AT THE KEYBOARD

Assume that you want to see two pieces of information for hourly workers: (1) the average wage rate, which you will call Average Rate in the output; and (2) 110 percent of the average rate, which you will call the Increased Rate. To get this information, you can make a calculated field in a new query from a Totals query. In other words, you use one query as a basis for another query.

Create the first query; you already know how to perform certain tasks for this query. The revised heading for the average rate will be Average Rate, so type *Average Rate: Wage Rate* in the Field cell. Note that you want the average of this field. Also, the grouping will be by the Salaried field. (To get hourly workers only, enter *Criteria: No*.) Confirm that your query resembles that in Figure B-31, then save the query and close it.

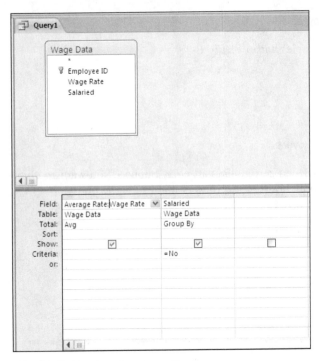

FIGURE B-31   A totals query with average

Now begin a new query. However, instead of bringing in a table to the query design, select a query. To start a new query, click the Create tab and then click the Query Design button in the Queries group. The Show Table dialog box appears. Click the Queries tab instead of using the default Tables tab, and select the query you just saved as a basis for the new query. The most difficult part of this query is to construct the expression for the calculated field. Conceptually, it is as follows:

Increased Rate: 1.1 * [The current average]

You use the new field name in the new query as the current average, and you treat the new name like a new field:

Increased Rate: 1.1 * [Average Rate]

The query within a query is shown in Figure B-32.

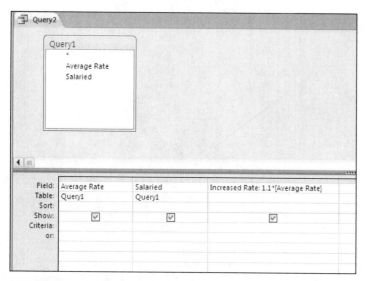

FIGURE B-32   A query within a query

Figure B-33 shows the output of the new query. Note that the calculated field is formatted.

FIGURE B-33   Output of an Expression in a Totals query

You do not need to print or save the query output, so return to Design view and close the query.

Using the Date() Function in Queries

Access has two important date function features:

- The built-in Date() function gives you today's date. You can use the function in query criteria or in a calculated field. The function "returns" the day on which the query is run; in other words, it inserts the value where the Date() function appears in an expression.
- Date arithmetic lets you subtract one date from another to obtain the difference—in number of days—between two calendar dates. For example, suppose you create the following expression:
  10/9/2012 – 10/4/2012
  Access would evaluate the expression as the integer 5 (9 minus 4 is 5).

As another example of how date arithmetic works, suppose you want to give each employee a one-dollar bonus for each day the employee has worked. You would need to calculate the number of days between the employee's date of hire and the day the query is run, and then multiply that number by $1.

You would find the number of elapsed days by using the following equation:

Date() – [Date Hired]

Also suppose that for each employee, you want to see the last name, employee ID, and bonus amount. You would set up the query as shown in Figure B-34.

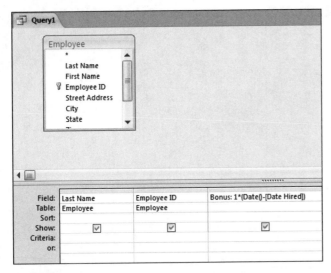

FIGURE B-34   Date arithmetic in a query

Assume that you set the format of the Bonus field to Currency. The output will be similar to that in Figure B-35, although your Bonus data will be different because you used a different date.

Last Name	Employee ID	Bonus
Brady	09911	$0.00
Howard	11411	$137.00
Smith	12345	$4,581.00
Smith	14890	$7,825.00
Jones	22282	$1,615.00
Ruth	71460	$3,411.00

FIGURE B-35   Output of query with date arithmetic

Using Time Arithmetic in Queries

Access also allows you to subtract the values of time fields to get an elapsed time. Assume that your database has a Job Assignments table showing the times that nonsalaried employees were at work during a day. The definition is shown in Figure B-36.

Field Name	Data Type
Employee ID	Text
ClockIn	Date/Time
ClockOut	Date/Time
DateWorked	Date/Time

FIGURE B-36   Date/Time data definition in the Job Assignments table

Assume that the DateWorked field is formatted for Long Date and that the ClockIn and ClockOut fields are formatted for Medium Time. Also assume that for a particular day, nonsalaried workers were scheduled as shown in Figure B-37.

Employee ID	ClockIn	ClockOut	DateWorked	Click to Add
09911	8:30:00 AM	4:30:00 PM	Friday, September 28, 2012	
11411	9:00:00 AM	3:00:00 PM	Friday, September 28, 2012	
14890	7:00:00 AM	5:00:00 PM	Friday, September 28, 2012	

FIGURE B-37   Display of date and time in a table

You want a query showing the elapsed time that your employees were on the premises for the day. When you add the tables, your screen may show the links differently. Click and drag the Job Assignments, Employee, and Wage Data table icons to look like those in Figure B-38.

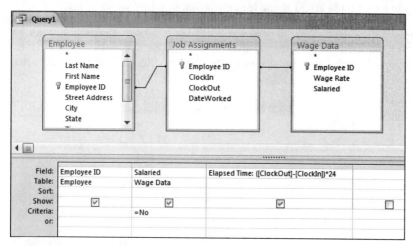

FIGURE B-38   Query setup for time arithmetic

Figure B-39 shows the output, which looks correct. For example, employee 09911 was at work from 8:30 a.m. to 4:30 p.m., which is eight hours. But how does the odd expression that follows yield the correct answers?

Employee ID	Salaried	Elapsed Time
09911	☐	8
11411	☐	6
14890	☐	10
*	☐	

FIGURE B-39   Query output for time arithmetic

([ClockOut] – [ClockIn]) * 24

Why wouldn't the following expression work?

[ClockOut] – [ClockIn]

Here is the answer: In Access, subtracting one time from the other yields the *decimal* portion of a 24-hour day. Returning to the example, you can see that employee 09911 worked eight hours, which is one-third of a day, so the time arithmetic function yields .3333. That is why you must multiply by 24—to convert from decimals to an hourly basis. Hence, for employee 09911, the expression performs the following calculation: $1/3 \times 24 = 8$.

Note that parentheses are needed to force Access to do the subtraction *first*, before the multiplication. Without parentheses, multiplication takes precedence over subtraction. For example, consider the following expression:

[ClockOut] – [ClockIn] * 24

In this example, ClockIn would be multiplied by 24, the resulting value would be subtracted from Clock-Out, and the output would be a nonsensical decimal number.

Deleting and Updating Queries

The queries presented in this tutorial so far have been Select queries. They select certain data from specific tables based on a given criterion. You also can create queries to update the original data in a database. Businesses use such queries often, and in real time. For example, when you order an item from a Web site, the company's database is updated to reflect your purchase through the deletion of that item from the company's inventory.

Consider an example. Suppose you want to give all nonsalaried workers a $0.50 per hour pay raise. Because you have only three nonsalaried workers, it would be easy to change the Wage Rate data in the table. However, if you had 3,000 nonsalaried employees, it would be much faster and more accurate to change the Wage Rate data by using an Update query that adds $0.50 to each nonsalaried employee's wage rate.

AT THE KEYBOARD

Now you will change each of the nonsalaried employees' pay via an Update query. Figure B-40 shows how to set up the query.

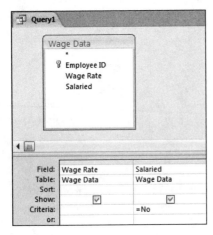

FIGURE B-40   Query setup for an Update query

So far, this query is just a Select query. Click the Update button in the Query Type group, as shown in Figure B-41.

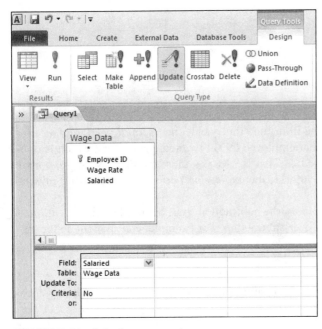

FIGURE B-41   Selecting a query type

Notice that you now have another line on the QBE grid called Update To:, which is where you specify the change or update the data. Notice that you will update only the nonsalaried workers by using a filter under the Salaried field. Update the Wage Rate data to Wage Rate plus $0.50, as shown in Figure B-42. Note that the update involves the use of brackets [ ], as in a calculated field.

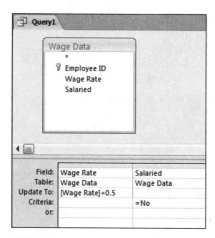

FIGURE B-42   Updating the wage rate for nonsalaried workers

Now run the query by clicking the Run button in the Results group. If you cannot run the query because it is blocked by Disabled Mode, click the Database Tools tab, then click Message Bar in the Show/Hide group. Click the Options button, choose "enable this content," and then click OK. When you successfully run the query, the warning message in Figure B-43 appears.

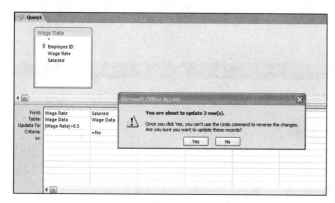

FIGURE B-43   Update query warning

When you click Yes, the records are updated. Check the updated records by viewing the Wage Data table. Each nonsalaried wage rate should be increased by $0.50. You could add or subtract data from another table as well. If you do, remember to put the field name in square brackets.

Another type of query is the Delete query, which works like Update queries. For example, assume that your company has been purchased by the state of Delaware, which has a policy of employing only state residents. Thus, you must delete (or fire) all employees who are not exclusively Delaware residents. To do that, you would create a Select query. Using the Employee table, you would click the Delete button in the Query Type group, then bring down the State field and filter only those records that were not in Delaware (DE). Do not perform the operation, but note that if you did, the setup would look like the one in Figure B-44.

FIGURE B-44   Deleting all employees who are not Delaware residents

Using Parameter Queries

A **Parameter query** is actually a type of Select query. For example, suppose your company has 5,000 employees and you want to query the database to find the same kind of information repeatedly, but about different employees each time. For example, you might want to know how many hours a particular employee has worked. You could run a query that you created and stored previously, but run it only for a particular employee.

AT THE KEYBOARD

Create a Select query with the format shown in Figure B-45.

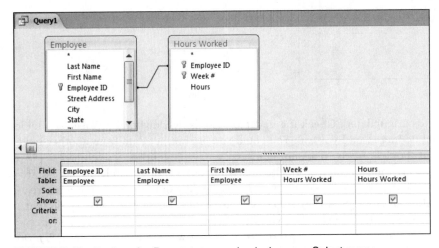

FIGURE B-45   Design of a Parameter query beginning as a Select query

In the Criteria line of the QBE grid for the Employee ID field, type what is shown in Figure B-46.

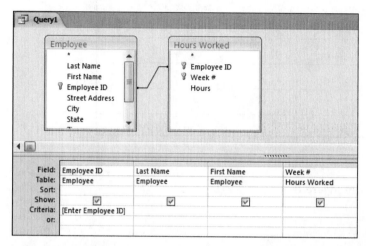

FIGURE B-46   Design of a Parameter query, continued

Note that the Criteria line uses square brackets, as you would expect to see in a calculated field. Now run the query. You will be prompted for the employee's ID number, as shown in Figure B-47.

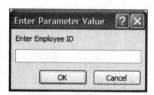

FIGURE B-47   Enter Parameter Value dialog box

Enter your own employee ID. Your query output should resemble that in Figure B-48.

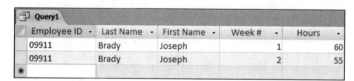

FIGURE B-48   Output of a Parameter query

MAKING SEVEN PRACTICE QUERIES

This portion of the tutorial gives you additional practice in creating queries. Before making these queries, you must create the specified tables and enter the records shown in the "Creating Tables" section of this tutorial. The output shown for the practice queries is based on those inputs.

AT THE KEYBOARD

For each query that follows, you are given a problem statement and a "scratch area." You also are shown what the query output should look like. Set up each query in Access and then run the query. When you are satisfied with the results, save the query and continue with the next one. Note that you will work with the Employee, Hours Worked, and Wage Data tables.

1.   Create a query that shows the employee ID, last name, state, and date hired for employees who live in Delaware *and* were hired after 12/31/99. Perform an ascending sort by employee ID. First click the Sort cell of the field, and then choose Ascending or Descending. Before creating your query, use the table shown in Figure B-49 to work out your QBE grid on paper.

Field					
Table					
Sort					
Show					
Criteria					
Or:					

FIGURE B-49   QBE grid template

Your output should resemble that in Figure B-50.

Query1

Employee ID ▾	Last Name ▾	State ▾	Date Hired ▾
11411	Howard	DE	8/1/2011
22282	Jones	DE	7/15/2004
*			

FIGURE B-50   Number 1 query output

2.   Create a query that shows the last name, first name, date hired, and state for employees who live in Delaware *or* were hired after 12/31/99. The primary sort (ascending) is on last name, and the secondary sort (ascending) is on first name. The Primary Sort field must be to the left of the Secondary Sort field in the query setup. Before creating your query, use the table shown in Figure B-51 to work out your QBE grid on paper.

Field				
Table				
Sort				
Show				
Criteria				
Or:				

FIGURE B-51   QBE grid template

If your name was Joe Brady, your output would look like that in Figure B-52.

Query2

Last Name ▾	First Name ▾	Date Hired ▾	State ▾
Brady	Joseph	9/14/2012	MD
Howard	Jane	8/1/2011	DE
Jones	Sue	7/15/2004	DE
Smith	Albert	7/15/1987	DE
Smith	John	6/1/1996	DE
*			

FIGURE B-52   Number 2 query output

3. Create a query that sums the number of hours worked by U.S. citizens and the number of hours worked by non-U.S. citizens. In other words, create two sums, grouped on citizenship. The heading for total hours worked should be Total Hours Worked. Before creating your query, use the table shown in Figure B-53 to work out your QBE grid on paper.

Field					
Table					
Total					
Sort					
Show					
Criteria					
Or:					

FIGURE B-53   QBE grid template

Your output should resemble that in Figure B-54.

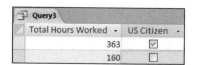

FIGURE B-54   Number 3 query output

4. Create a query that shows the wages owed to hourly workers for Week 1. The heading for the wages owed should be Total Owed. The output headings should be Last Name, Employee ID, Week #, and Total Owed. Before creating your query, use the table shown in Figure B-55 to work out your QBE grid on paper.

Field					
Table					
Sort					
Show					
Criteria					
Or:					

FIGURE B-55   QBE grid template

If your name was Joe Brady, your output would look like that in Figure B-56.

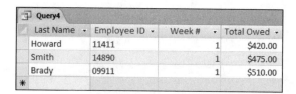

FIGURE B-56   Number 4 query output

5. Create a query that shows the last name, employee ID, hours worked, and overtime amount owed for hourly employees who earned overtime during Week 2. Overtime is paid at 1.5 times the normal hourly rate for all hours worked over 40. Note that the amount shown in the query should be just the overtime portion of the wages paid. Also, this is not a Totals query—amounts should be shown for individual workers. Before creating your query, use the table shown in Figure B-57 to work out your QBE grid on paper.

Field					
Table					
Sort					
Show					
Criteria					
Or:					

FIGURE B-57   QBE grid template

If your name was Joe Brady, your output would look like that in Figure B-58.

Last Name	Employee ID	Hours	OT Pay
Howard	11411	50	$157.50
Brady	09911	55	$191.25

FIGURE B-58   Number 5 query output

6. Create a Parameter query that shows the hours employees have worked. Have the Parameter query prompt for the week number. The output headings should be Last Name, First Name, Week #, and Hours. This query is for nonsalaried workers only. Before creating your query, use the table shown in Figure B-59 to work out your QBE grid on paper.

Field					
Table					
Sort					
Show					
Criteria					
Or:					

FIGURE B-59   QBE grid template

Run the query and enter 2 when prompted for the week number. Your output should look like that in Figure B-60.

Last Name	First Name	Week #	Hours
Howard	Jane	2	50
Smith	Albert	2	40
Brady	Joseph	2	55

FIGURE B-60   Number 6 query output

7. Create an Update query that gives certain workers a merit raise. First, you must create an additional table, as shown in Figure B-61.

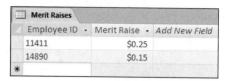

Merit Raises		
Employee ID ▾	Merit Raise ▾	Add New Field
11411	$0.25	
14890	$0.15	
*		

FIGURE B-61   Merit Raises table

8. Create a query that adds the Merit Raise to the current Wage Rate for employees who will receive a raise. When you run the query, you should be prompted with *You are about to update two rows.* Check the original Wage Data table to confirm the update. Before creating your query, use the table shown in Figure B-62 to work out your QBE grid on paper.

Field					
Table					
Update to					
Criteria					
Or:					

FIGURE B-62   QBE grid template

CREATING REPORTS

Database packages let you make attractive management reports from a table's records or from a query's output. If you are making a report from a table, the Access report generator looks up the data in the table and puts it into report format. If you are making a report from a query's output, Access runs the query in the background (you do not control it or see it happen) and then puts the output in report format.

There are different ways to make a report. One method is to create one from scratch in Design view, but this tedious process is not explained in this tutorial. A simpler way is to select the query or table on which the report is based and then click Create Report. This streamlined method of creating reports is explained in this tutorial.

Creating a Grouped Report

This tutorial assumes that you already know how to create a basic ungrouped report, so this section teaches you how to make a grouped report. If you do not know how to create an ungrouped report, you can learn by following the first example in the upcoming section.

AT THE KEYBOARD

Suppose you want to create a report from the Hours Worked table. Select the table by clicking it once. Click the Create tab, then click Report in the Reports group. A report appears, as shown in Figure B-63.

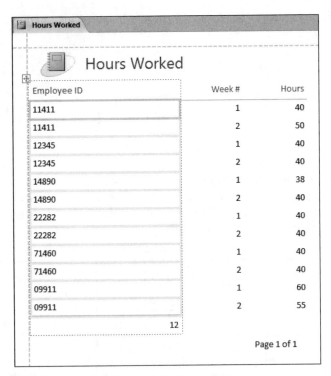

FIGURE B-63   Initial report based on a table

On the Design tab, select the Group and Sort button in the Grouping and Totals group. Your report will have an additional selection at the bottom, as shown in Figure B-64.

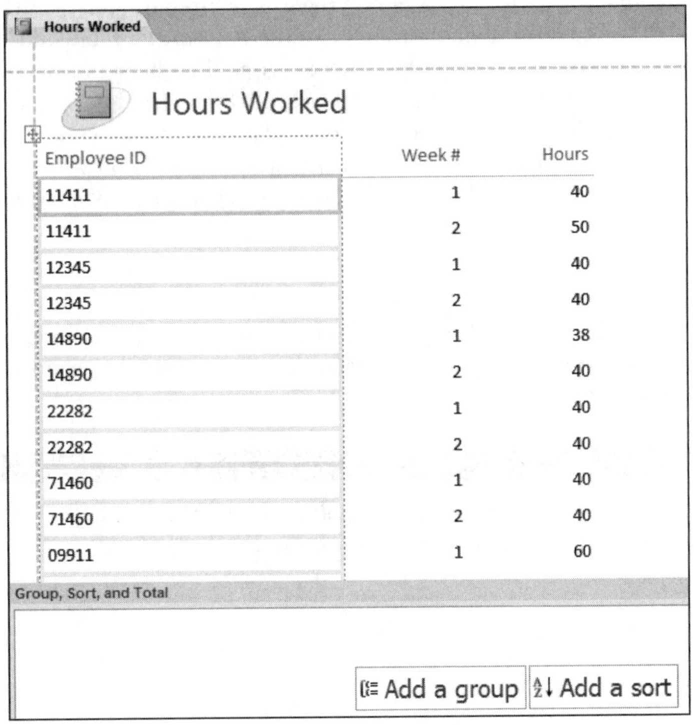

FIGURE B-64   Report with Grouping and Sorting options

Click the Add a group button at the bottom of the report, and then select Employee ID. Your report will be grouped as shown in Figure B-65.

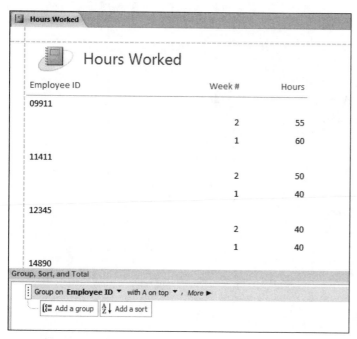

FIGURE B-65 Grouped report

To complete this report, you need to total the hours for each employee by selecting the Hours column heading. Your report will show that the entire column is selected. On the Design tab, click the Totals button in the Grouping and Totals group, and then choose Sum from the menu, as shown in Figure B-66.

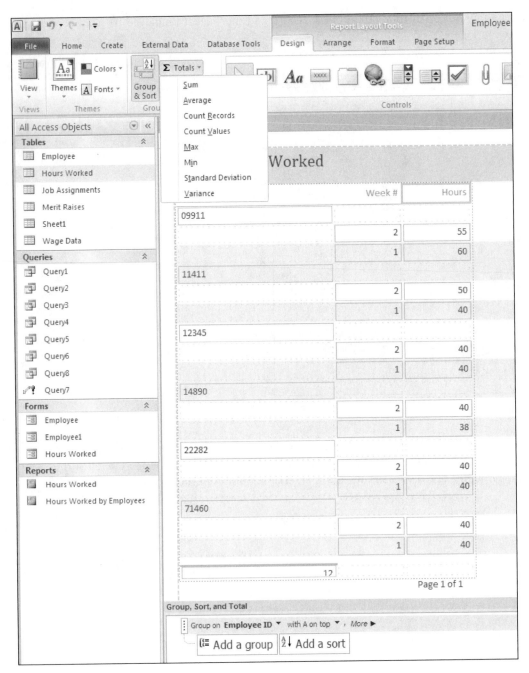

FIGURE B-66   Hours column selected

Your report will look like the one in Figure B-67.

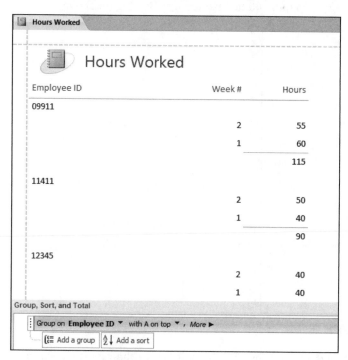

FIGURE B-67 Completed report

Your report is currently in Layout view. To see how the final report looks when printed, click the Design tab and select Report view from the Views group. Your report looks like the one in Figure B-68, although only a portion is shown in the figure.

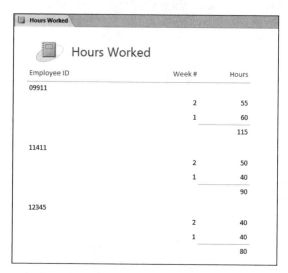

FIGURE B-68 Report in Report view

NOTE

To change the picture or logo in the upper-left corner of the report, click the notebook symbol and press the Delete key. You can insert a logo in place of the notebook by clicking the Design tab and then clicking Logo in the Controls group.

Moving Fields in Layout View

If you group records based on more than one field in a report, the report will have an odd "staircase" look or display repeated data, or it will have both problems. Next, you will learn how to overcome these problems in Layout view.

Suppose you make a query that shows an employee's last name, first name, week number, and hours worked, and then you make a report from that query, grouping on last name only. See Figure B-69.

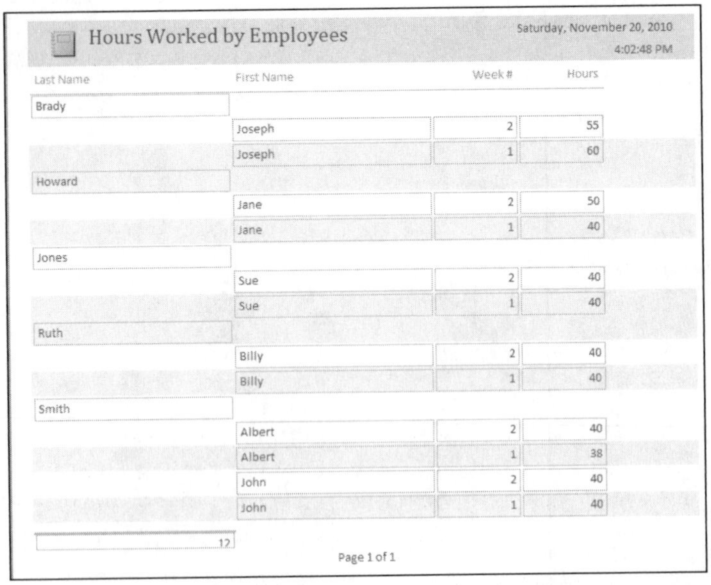

FIGURE B-69   Query-based report grouped on last name

As you preview the report, notice the repeating data from the First Name field. In the report shown in Figure B-69, notice that the first name repeats for each week worked—hence, the staircase effect. The Week # and Hours fields are shown as subordinate to Last Name, as desired.

Suppose you want the last name and first name to appear on the same line. If so, take the report into Layout view for editing. Click the first record for the First Name (in this case, Joseph), and drag the name up to the same line as the Last Name (in this case, Brady). Your report will now show the First Name on the same line as Last Name, thereby eliminating the staircase look, as shown in Figure B-70.

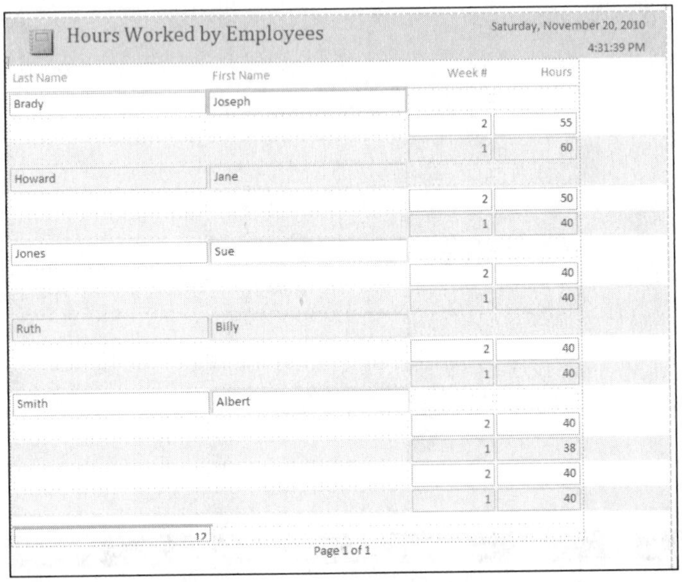

FIGURE B-70   Report in Layout view with Last Name and First Name on the same line

You can now add the sum of Hours for each group. Also, if you want to add more fields to your report, such as Street Address and Zip, you can repeat the preceding procedure.

IMPORTING DATA

Text or spreadsheet data is easy to import into Access. In business, it is often necessary to import data because companies use disparate systems. For example, assume that your healthcare coverage data is on the human resources manager's computer in a Microsoft Excel spreadsheet. Open the Excel application and then create a spreadsheet using the data shown in Figure B-71.

	A	B	C
1	Employee ID	Provider	Level
2	11411	BlueCross	family
3	12345	BlueCross	family
4	14890	Coventry	spouse
5	22282	None	none
6	71460	Coventry	single
7	09911	BlueCross	single

FIGURE B-71   Excel data

Save the file and then close it. Now you can easily import the spreadsheet data into a new table in Access. With your Employee database open, click the External Data tab, then click Excel in the Import & Link group. Browse to find the Excel file you just created, and make sure the first radio button is selected to import the source data into a new table in the current database (see Figure B-72). Click OK.

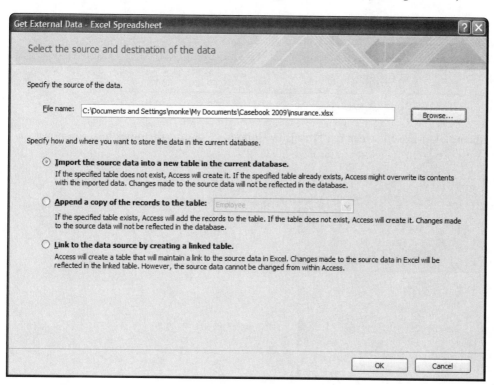

FIGURE B-72   Importing Excel data into a new table

Choose the correct worksheet. Assuming that you have just one worksheet in your Excel file, your next screen should look like the one in Figure B-73.

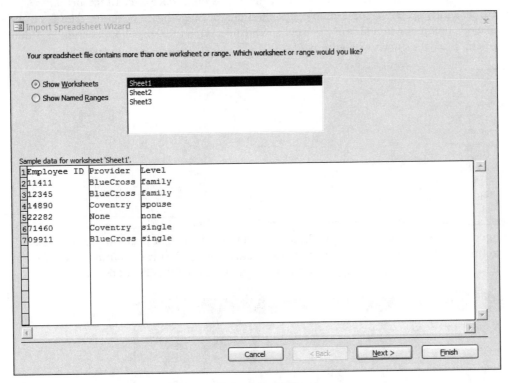

FIGURE B-73   First screen in the Import Spreadsheet Wizard

Choose Next, and then make sure to select the First Row Contains Column Headings box, as shown in Figure B-74.

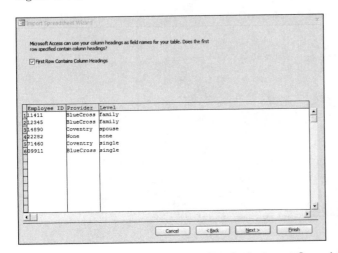

FIGURE B-74   Choosing column headings in the Import Spreadsheet Wizard

Choose Next. Accept the default setting for each field you are importing on the screen. Each field is assigned a text data type, which is correct for this table. Your screen should look like the one in Figure B-75.

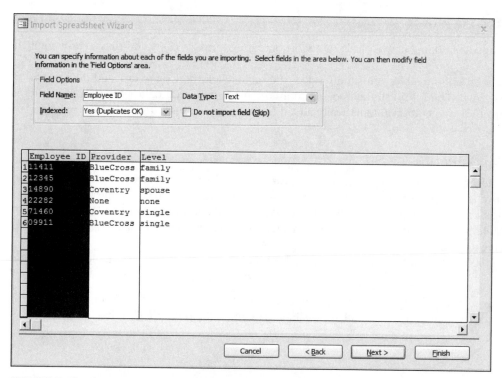

FIGURE B-75 Choosing the data type for each field in the Import Spreadsheet Wizard

Choose Next. In the next screen of the wizard, you will be prompted to create an index—that is, to define a primary key. Because you will store your data in a new table, choose your own primary key (Employee ID), as shown in Figure B-76.

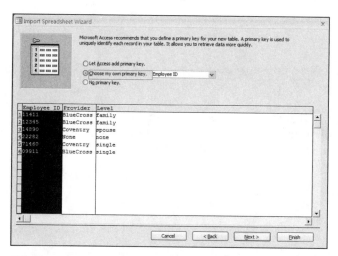

FIGURE B-76 Choosing a primary key field in the Import Spreadsheet Wizard

Continue through the wizard, giving your table an appropriate name. After importing the table, take a look at its design by highlighting the Table option and clicking the Design button. Note that each field is very wide. Adjust the field properties as needed.

MAKING FORMS

Forms simplify the process of adding new records to a table. Creating forms is easy, and they can be applied to one or more tables.

When you base a form on one table, you simply select the table, click the Create tab, and then select Form from the Forms group. The form will then contain only the fields from that table. When data is entered into the form, a complete new record is automatically added to the table. Forms with two tables are discussed next.

Making Forms with Subforms

You also can create a form that contains a subform, which can be useful when the form is based on two or more tables. Return to the example Employee database to see how forms and subforms would be useful for viewing all of the hours that each employee worked each week. Suppose you want to show all of the fields from the Employee table; you also want to show the hours each employee worked by including all fields from the Hours Worked table as well.

To create the form and subform, first create a simple one-table form on the Employee table. Follow these steps:

1. Click once to select the Employee table. Click the Create tab, then click Form in the Forms group. After the main form is complete, it should resemble the one in Figure B-77.

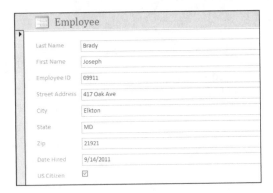

FIGURE B-77   The Employee form

2. To add the subform, take the form into Design view. On the Design tab, make sure that the Use Control Wizards option is selected, scroll to the bottom row of buttons in the Controls group, and click the Subform/Subreport button, as shown in Figure B-78.

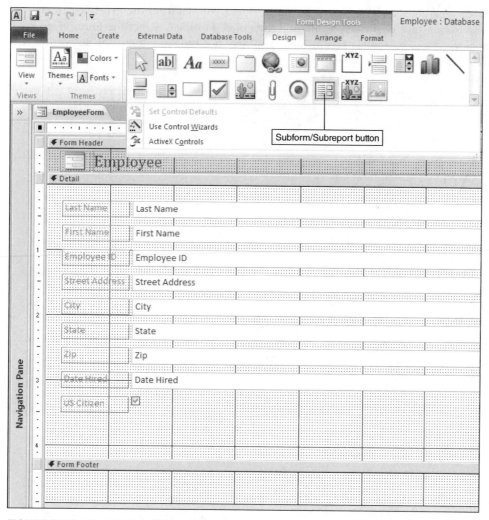

FIGURE B-78   The Subform/Subreport button

3.   Use your cursor to stretch out the box under your main form. The dialog box shown in Figure B-79 appears.

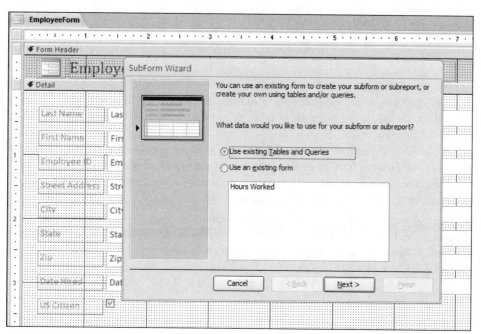

FIGURE B-79   Adding a subform

4. Select Use existing Tables and Queries, then select Hours Worked from the list. Click Next, select Choose from a List, click Next again, and then click Finished. Select the Form view. Your form and subform should resemble Figure B-80. You may need to stretch out the subform box in Design view if all fields are not visible.

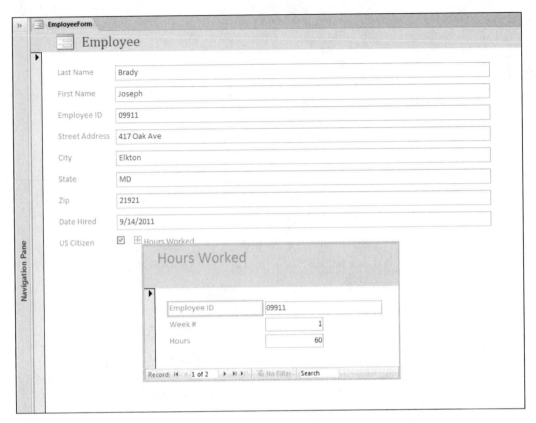

FIGURE B-80   Form with subform

TROUBLESHOOTING COMMON PROBLEMS

Access is a powerful program, but it is complex and sometimes difficult for new users. People sometimes unintentionally create databases that have problems. Some of these common problems are described below, along with their causes and corrections.

1. *"I saved my database file, but I can't find it on my computer or my external secondary storage medium! Where is it?"*

    You saved your file to a fixed disk or a location other than the Documents folder. Click the Windows Start button, then use the Search option to find all files ending in .accdb (search for *.accdb). If you saved the file, it is on the hard drive (C:\) or a network drive. Your site assistant can tell you the drive designators.

2. *"What is a 'duplicate key field value'? I'm trying to enter records into my Sales table. The first record was for a sale of product X to customer 101, and I was able to enter that one. But when I try to enter a second sale for customer #101, Access tells me I already have a record with that key field value. Am I allowed to enter only one sale per customer?"*

    Your primary key field needs work. You may need a compound primary key—a combination of the customer number and some other field(s). In this case, the customer number, product number, and date of sale might provide a unique combination of values, or you might consider using an invoice number field as a key.

3. *"My query reads 'Enter Parameter Value' when I run it. What is that?"*

This problem almost always indicates that you have misspelled a field name in an expression in a Criteria field or calculated field. Access is very fussy about spelling; for example, it is case-sensitive. Access is also "space-sensitive," meaning that when you insert a space in a field name when defining a table, you must also include a space in the field name when you reference it in a query expression. Fix the typo in the query expression.

4. *"I'm getting an enormous number of rows in my query output—many times more than I need. Most of the rows are duplicates!"*

This problem is usually caused by a failure to link all of the tables you brought into the top half of the query generator. The solution is to use the manual click-and-drag method to link the common fields between tables. The spelling of the field names is irrelevant because the link fields need not have the same spelling.

5. *"For the most part, my query output is what I expected, but I am getting one or two duplicate rows or not enough rows."*

You may have linked too many fields between tables. Usually, only a single link is needed between two tables. It is unnecessary to link each common field in all combinations of tables; it is usually sufficient to link the primary keys. A simplistic explanation for why overlinking causes problems is that it causes Access to "overthink" and repeat itself in its answer.

On the other hand, you might be using too many tables in the query design. For example, you brought in a table, linked it on a common field with some other table, but then did not use the table. In other words, you brought down none of its fields, and/or you used none of its fields in query expressions. In this case, if you got rid of the table, the query would still work. Click the unneeded table's header at the top of the QBE area, and press the Delete key to see if you can make the few duplicate rows disappear.

6. *"I expected six rows in my query output, but I got only five. What happened to the other one?"*

Usually, this problem indicates a data entry error in your tables. When you link the proper tables and fields to make the query, remember that the linking operation joins records from the tables *on common values* (*equal* values in the two tables). For example, if a primary key in one table has the value "123," the primary key or the linking field in the other table should be the same to allow linking. Note that the text string "123" is not the same as the text string "123 "—the space in the second string is considered a character too. Access does not see unequal values as an error. Instead, Access moves on to consider the rest of the records in the table for linking. The solution is to examine the values entered into the linked fields in each table and fix any data entry errors.

7. *"I linked fields correctly in a query, but I'm getting the empty set in the output. All I get are the field name headings!"*

You probably have zero common (equal) values in the linked fields. For example, suppose you are linking on Part Number, which you declared as text. In one field, you have part numbers "001", "002", and "003"; in the other table, you have part numbers "0001," "0002," and "0003." Your tables have no common values, which means that no records are selected for output. You must change the values in one of the tables.

8. *"I'm trying to count the number of today's sales orders. A Totals query is called for. Sales are denoted by an invoice number, and I made that a text field in the table design. However, when I ask the Totals query to 'Sum' the number of invoice numbers, Access tells me I cannot add them up! What is the problem?"*

Text variables are words! You cannot add words, but you can count them. Use the Count Totals operator (not the Sum operator) to count the number of sales, each being denoted by an invoice number.

9. *"I'm doing time arithmetic in a calculated field expression. I subtracted the Time In from the Time Out and got a decimal number! I expected eight hours, and I got the number .33333. Why?"*

[Time Out] – [Time In] yields the decimal percentage of a 24-hour day. In your case, eight hours is one-third of a day. You must complete the expression by multiplying by 24: ([Time Out] – [Time In]) * 24. Don't forget the parentheses.

10. *"I formatted a calculated field for Currency in the query generator, and the values did show as currency in the query output; however, the report based on the query output does not show the dollar sign in its output. What happened?"*

Go to the report Design view. A box in one of the panels represents the calculated field's value. Click the box and drag to widen it. That should give Access enough room to show the dollar sign as well as the number in the output.

11. *"I told the Report Wizard to fit all of my output to one page. It does print to one page, but some of the data is missing. What happened?"*

Access fits all the output on one page by leaving data out. If you can tolerate having the output on more than one page, deselect the Fit to a Page option in the wizard. One way to tighten output is to enter Design view and remove space from each box that represents output values and labels. Access usually provides more space than needed.

12. *"I grouped on three fields in the Report Wizard, and the wizard prints the output in a staircase fashion. I want the grouping fields to be on one line. How can I do that?"*

Make adjustments in Design view and Layout view. See the "Creating Reports" section of this tutorial for instructions on making these adjustments.

13. *"When I create an Update query, Access tells me that zero rows are updating or more rows are updating than I want. What is wrong?"*

If your Update query is not set up correctly (for example, if the tables are not joined properly), Access will either try not to update anything, or it will update all of the records. Check the query, make corrections, and run it again.

14. *"I made a Totals query with a Sum in the Group By row and saved the query. Now when I go back to it, the Sum field reads 'Expression,' and 'Sum' is entered in the field name box. Is that wrong?"*

Access sometimes changes the Sum field when the query is saved. The data remains the same, and you can be assured your query is correct.

15. *"I cannot run my Update query, but I know it is set up correctly. What is wrong?"*

Check the security content of the database by clicking the Security Content button. You may need to enable certain actions.

PRELIMINARY CASE: THE FIREPLACE SHOP

Setting up a Relational Database to Create Tables, Forms, Queries, and Reports

PREVIEW

In this case, you will create a relational database for a business that sells fireplaces and heating stoves. First, you will create four tables and populate them with data. Next, you will create a form and subform for recording customers and their orders. You will create five queries: a parameter query, a select query, an update query, a totals query, and a query used as the basis for a report. You will create the report from the fifth query.

PREPARATION

- Before attempting this case, you should have some experience using Microsoft Access.
- Complete any part of Tutorial B that your instructor assigns, or refer to the tutorial as necessary.

BACKGROUND

Your aunt and uncle own a store in Atlanta called Family Fireplaces (FF) that sells fireplaces and heating stoves. You might think that the South is warm throughout the year, but winters there can be cool and most houses require heating. However, insulation in many houses is minimal, resulting in large heating bills when the temperature drops. In economic hard times, people are interested in reducing their heating bills. FF offers alternative heating sources such as gas fireplace inserts, wood-burning stoves, and gas stoves to keep homes warm in the winter.

Because you are proficient in Access, your aunt and uncle have asked you to help them complete a database that tracks customer orders and installations. Currently, the company tracks this information manually. Your aunt and uncle hired a summer intern last year to do the database design. Your job is to implement the database and create several outputs to help run the business.

Your first job is to create the tables and populate them with data. The database design includes four tables, as shown in Figures 1-1, 1-2, 1-3, and 1-4:

- Customers, which keeps track of each customer's ID number, name, address, e-mail address, telephone numbers, and credit card number for billing purposes
- Orders, which keeps track of each order number, customer ID number, product number, date purchased, and whether the product has been installed
- Products, which lists each product ID number, description, manufacturer, BTUs (a measure of heat output), and price
- Work Orders, which keeps track of each work order ID number, the technician who does the job, order number, date installed, and start and end times for installing the product

As the owners, your aunt and uncle have a few requirements for information output in the database besides simply recording the data. First, they want you to create an easier way to record a sale using the customer's information and order data. You can accomplish this task by creating a form and subform.

One of the company's suppliers has had a recall of its fireplaces and stoves. The owners want to e-mail and call all customers who have bought the recalled products. You explain that a parameter query would produce the list of customers, and that you can build the query to prompt for a manufacturer. The query can then be used again if another recall occurs, or if customers who own products from a certain manufacturer need to be contacted before a sale or for some other reason.

Your aunt and uncle also want to be able to respond quickly when people call with questions about which stove or fireplace will fit their needs. For example, a caller recently asked which products had more than 30,000 BTUs but were priced under $2,000. Again, you explain that a well-designed query could help answer customer questions.

To help marketing efforts, the owners always want to know which products sell the best. They believe that the most popular products are not necessarily based on price; sometimes their higher-priced or medium-priced products are the most popular. Therefore, the owners would like a listing of which products sell the most and which sell the least.

Also, manufacturers sometimes raise their prices, which means that FF must increase its prices to customers as well. The owners would like an easy way to change prices for all products from a specified manufacturer. You suggest using an update query to adjust prices.

Finally, your aunt and uncle would like an attractive report that lists the amounts FF owes the two company technicians who install fireplaces and stoves for customers. The two technicians, Frank and Sam, are paid by the hour for installations.

ASSIGNMENT 1: CREATING TABLES

Use Access to create tables that contain the fields shown in Figures 1-1 through 1-4; these tables were discussed in the Background section. Populate the database tables as shown. Add your name to the Customers table with an appropriate customer ID; complete the entry by adding your address, phone numbers, e-mail address, and a fictional credit card number.

The database contains the following four tables:

Customers											
Customer ID	Last Name	First Name	Address	City	State	Zip	Telephone	Cell Phone	Email Address	Credit Card
B-17	Bianco	Anna	9 Pleasant Way	Atlanta	GA	30600	404-887-4673	404-876-3376	aberry@hotmail.com	443376562837
F-59	Franklin	Gina	1012 Peachtree St	Atlanta	GA	30600	404-887-2342	404-765-1263	gf59@gmail.com	443398764532
L-29	LaGoia	Maria	5490 West 5th	Atlanta	GA	30600	404-234-8876	569-001-0989	mrl@hotmail.com	443352635423
M-62	McMillan	Annabelle	59 W. Central Ave	Atlanta	GA	30600	404-998-3928	404-887-3829	belle@comcast.net	443355463212
P-91	Prince	Jill	89 Orchard	Atlanta	GA	30600	404-887-9238	404-342-9087	pao@comcast.net	443367256543
Q-13	Queller	Sally	54 Oak Ave	Atlanta	GA	30600	404-987-3427	569-984-3894	quinn45@gmail.com	443398765439
S-63	Sandy	Patricia	1700 E. Lincoln Ave	Atlanta	GA	30600	404-765-3342	404-121-4736	patti1@gmail.com	443398762534
Z-30	Zepher	Joan	58 W. Central Ave	Atlanta	GA	30600	404-675-0091	404-776-4536	zern@comcast.net	443357643254

FIGURE 1-1 The Customers table

Orders					
Order Number	Customer ID	Product ID	Date Purchased	Installed?	Cl
1	Z-30	104	8/12/2012	✓	
10	Q-13	101	12/1/2012	☐	
2	S-63	101	9/15/2012	✓	
3	Q-13	103	9/18/2012	✓	
4	P-91	102	10/1/2012	✓	
5	M-62	101	10/12/2012	✓	
6	L-29	101	10/20/2012	✓	
7	F-59	104	11/12/2012	☐	
8	B-17	105	11/13/2012	☐	
9	S-63	101	11/23/2012	☐	
*				☐	

FIGURE 1-2 The Orders table

Products					
Product ID ▾	Description ▾	Manufacturer ▾	BTUs ▾	Price ▾	Click
101	Gas Fireplace	Majestic	25000	$2,130.00	
102	Gas Stove - White	Napoleon	23000	$1,290.00	
103	Gas Stove - Black	Napoleon	23000	$1,280.00	
104	Gas Stove - Circular	Imperial	60000	$3,250.00	
105	Wood Stove - Grey	Mansfield	80000	$1,890.00	
*					

FIGURE 1-3   The Products table

Work Orders						
Work Order ▾	Technician ▾	Order Number ▾	Date Installed ▾	Start Time ▾	End Time ▾	Click to
1025	Sam	1	9/1/2012	8:00:00 AM	4:00:00 PM	
1026	Frank	2	9/1/2012	7:30:00 AM	2:00:00 PM	
1027	Sam	3	10/2/2012	10:00:00 AM	6:00:00 PM	
1028	Sam	4	10/12/2012	10:00:00 AM	5:00:00 PM	
1029	Sam	5	10/30/2012	7:30:00 AM	4:30:00 PM	
1030	Frank	6	10/30/2012	11:00:00 AM	7:00:00 PM	
*						

FIGURE 1-4   The Work Orders table

ASSIGNMENT 2: CREATING A FORM, QUERIES, AND A REPORT

Assignment 2A: Creating a Form

Create a form for easy recording of customers and their orders. The main form should be based on the Customers table, and the subform should be inserted with the fields from the Orders table. Save the form as Customers. View one record; if required by your instructor, print the record. Your output should resemble that in Figure 1-5.

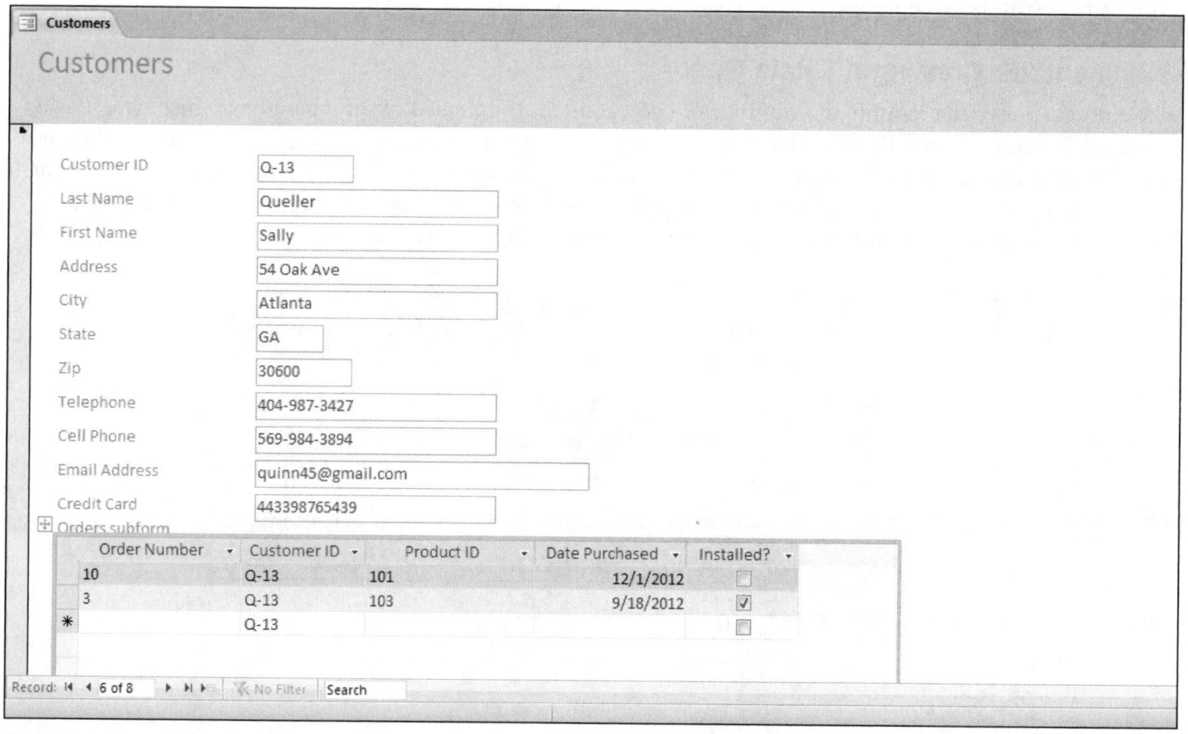

Customers	
Customers	
Customer ID	Q-13
Last Name	Queller
First Name	Sally
Address	54 Oak Ave
City	Atlanta
State	GA
Zip	30600
Telephone	404-987-3427
Cell Phone	569-984-3894
Email Address	quinn45@gmail.com
Credit Card	443398765439

Orders subform

Order Number ▾	Customer ID ▾	Product ID ▾	Date Purchased ▾	Installed? ▾
10	Q-13	101	12/1/2012	☐
3	Q-13	103	9/18/2012	☑
*	Q-13			☐

Record: I◄ ◄ 6 of 8 ► ►I ►* 🔾 No Filter   Search

FIGURE 1-5   The Customers form with subform

Assignment 2B: Creating a Parameter Query

Create a parameter query that prompts for a manufacturer and then displays columns for customers' Last Name, First Name, Telephone, Cell Phone, and Email Address. Run the query for all customers who bought products manufactured by the Napoleon Company. Save your query as Customers of Specified Manufacturer. Your output should resemble Figure 1-6 when you enter Napoleon at the prompt.

Customers of Specified Manufacturer				
Last Name ▾	First Name ▾	Telephone ▾	Cell Phone ▾	Email Address ▾
Prince	Jill	404-887-9238	404-342-9087	pao@comcast.net
Queller	Sally	404-987-3427	569-984-3894	quinn45@gmail.com
*				

FIGURE 1-6   Customers of Specified Manufacturer query

Run the query. Print the results if required.

Assignment 2C: Creating a Select Query

Create a query to display all products that have more than 30,000 BTUs and that cost less than $2,000. Save your query as High BTU and Low Price. Your output should resemble that in Figure 1-7. Print the output if desired.

High BTU and Low Price			
Description ▾	Manufacturer ▾	BTUs ▾	Price ▾
Wood Stove - Grey	Mansfield	80000	$1,890.00
*			

FIGURE 1-7   High BTU and Low Price query

Assignment 2D: Creating an Update Query

Create a query that increases the price of all products manufactured by the Majestic Company by 10%. Run the query to test it. Save the query as Increased Price on Majestic.

Assignment 2E: Creating a Totals Query

Create a totals query that counts the number of each product purchased by FF customers. Display columns for Number Purchased, Description, Manufacturer, and Price. (Note that the Number Purchased heading is a column heading change from the default setting provided by the query generator.) Sort your output so that products are listed from most popular to least popular. Save your query as Most Popular Products. Your output should resemble that in Figure 1-8. Print the output if instructed to do so.

Most Popular Products			
Number Purchased ▾	Description ▾	Manufacturer ▾	Price ▾
5	Gas Fireplace	Majestic	$2,130.00
2	Gas Stove - Circular	Imperial	$3,250.00
1	Wood Stove - Grey	Mansfield	$1,890.00
1	Gas Stove - White	Napoleon	$1,290.00
1	Gas Stove - Black	Napoleon	$1,280.00

FIGURE 1-8   Most Popular Products query

Assignment 2F: Generating a Report

Generate a report named Technician Pay based on a query. The query should display fields for Technician, Order Number, Date Installed, and Pay, which is a calculated field. Save the query as For Report. From that query, create a report that lists the total pay owed to each technician and displays a grand total at the bottom. Make sure that all fields and data are visible. Your report output should resemble that in Figure 1-9.

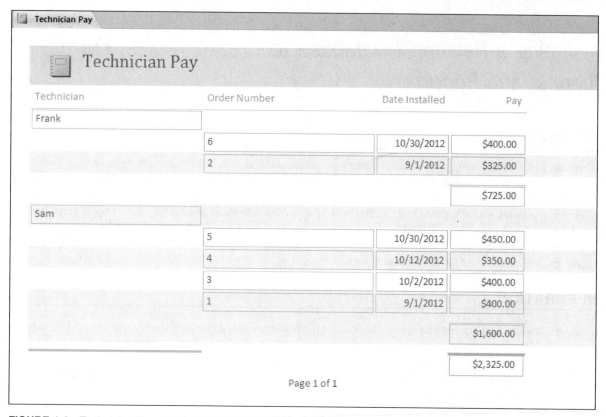

FIGURE 1-9 Technician Pay report

If you are working with a portable storage disk or USB key, make sure that you remove it *after* closing the database file.

DELIVERABLES

Assemble the following deliverables for your instructor, either electronically or in printed form:

1. Four tables
2. Form and subform: Customers
3. Query 1: Customers of Specified Manufacturer
4. Query 2: High BTU and Low Price
5. Query 3: Increased Price on Majestic
6. Query 4: Most Popular Products
7. Query 5: For Report
8. Report: Technician Pay
9. Any other required printouts or electronic media

Staple all the pages together. Include your name and class number at the top of the page. If required, make sure that your electronic media are labeled.

DECORATIVE CONCRETE DATABASE

Designing a Relational Database to Create Tables, Forms, Queries, and Reports

PREVIEW

In this case, you will design a relational database for a company that sells and installs decorative concrete. After your design is completed and correct, you will create database tables and populate them with data. You will produce one form with a subform that allows you to record jobs for the concrete company, and you will produce five queries and a report. The queries will address the following questions: Which customers have a pool and more than two acres of land? Which jobs are booked for a particular month and business partner? What type of concrete does the company use most? What is the most lucrative product? Your report, based on a query, will display customer payments to the company.

PREPARATION

- Before attempting this case, you should have some experience in database design and in using Microsoft Access.
- Complete any part of Tutorial A that your instructor assigns.
- Complete any part of Tutorial B that your instructor assigns, or refer to the tutorial as necessary.
- Refer to Tutorial F as necessary.

BACKGROUND

Two of your college friends have started a business to sell and install decorative concrete. Carl, whose father owns a concrete business, has extensive experience in pouring and laying concrete driveways and foundations. Candy, an art major, has done extensive decorative tiling and mosaics for friends. The couple decided to form a company called C&C Concrete. Decorative concrete is a popular alternative to tiling and plain concrete. Patterns can be stamped into wet concrete to make it look like expensive brick, and colors and alternative textures can be added to create further effects. Customers use decorative concrete for pool areas, sidewalks, paths, countertops, and flooring.

C&C Concrete has grown rapidly, and Candy and Carl cannot keep up with their scheduled jobs without a computer system. They know you have experience in database design and implementation, so they have asked for your help. Customer billing and bill collection are handled by a third party, so the database does not have to account for them.

Your job is to design a database for jobs and scheduling, and then to implement the database. Before you begin, however, you need to know a few details about the business. Candy and Carl value their customers and keep detailed records about them. Candy and Carl like to get to know each customer personally, which enables more targeted marketing. For example, new customers fill out a survey that records their name and address, number of acres on their property, whether they have a pool, and garage size (one, two, or three cars).

C&C Concrete sells three types of decorative concrete, as shown in Figure 2-1.

Type	Description	Price per Square Foot
Basic	Basic stamped one color	$10
Upgraded	Stamped with two to three colors and border	$16
Superior	Upgraded with hand-applied designs	$20

FIGURE 2-1 Types of concrete

The basic type of decorative concrete resembles brick; this colored concrete is poured and stamped. The upgraded concrete product features additional pattern coloring and costs more per square foot. The superior product features embellishments and hand applications.

When Candy and Carl get a new job with a customer, they record necessary details such as the type of decorative concrete ordered, the amount of square footage covered, and the address where the work will be performed. This address is not necessarily the customer's home address; jobs could be completed at a ski house in the mountains or at a summer house on a lake. To get to know customers better, Candy and Carl like to record as many additional comments as possible. Having an easy way to record this information, such as in a form, would be useful to their business and for future marketing campaigns. In addition, the owners want to know how many jobs each type of concrete generates. They also want to know which product is the most lucrative for the company.

The two owners are very busy and they don't employ extra help, so they need to schedule and track projects carefully to avoid overbooking. Therefore, your database should be able to record the schedule of each customer job and indicate which of the two owners will do the job. Keep in mind that all jobs take longer than one day to complete.

Finally, although billing and payments are handled by a third-party company, Candy and Carl would like to know what each customer owes. You will create a report that summarizes information about customer payments.

ASSIGNMENT 1: CREATING THE DATABASE DESIGN

In this assignment, you will design your database tables using a word-processing program. Pay close attention to the logic and structure of the tables. Do not start developing your Access database in Assignment 2 before getting feedback from your instructor on Assignment 1. Keep in mind that you will need to examine the requirements in Assignment 2 to design your fields and tables properly. It is good programming practice to look at the required outputs before beginning your design. When designing the database, observe the following guidelines:

- First, determine the tables you will need by listing the name of each table and the fields it should contain. Avoid data redundancy. Do not create a field if it can be created by a "calculated field" in a query.
- You will need transaction tables. Think about what business events occur with each customer's jobs. Avoid duplicating data.
- Document your tables using the table feature of your word processor. Your tables should resemble the format shown in Figure 2-2.
- You must mark the appropriate key field(s) by entering an asterisk (*) next to the field name.
- Print the database design if your instructor requires it.

Table Name	
Field Name	Data Type (text, numeric, currency, etc.)
...	...
...	...

FIGURE 2-2 Table design

N O T E

Have your design approved before beginning Assignment 2; otherwise, you may need to redo Assignment 2.

ASSIGNMENT 2: CREATING THE DATABASE, QUERIES, AND REPORT

In this assignment, you will first create database tables in Access and populate them with data. Next, you will create a form, queries, and a report.

Assignment 2A: Creating Tables in Access

In this part of the assignment, you will create your tables in Access. Use the following guidelines:

- Create at least 10 customers and at least 15 concrete installation jobs.
- Use the types of concrete and their descriptions and prices listed in Figure 2-1.
- Appropriately limit the size of the text fields; for example, a telephone number does not need the default length of 255 characters.
- Print all tables if your instructor requires it.

Assignment 2B: Creating Forms, Queries, and a Report

You will generate one form with a subform, five queries, and one report, as outlined in the Background section of this case.

Form

Create a form and subform based on your Customer table and Jobs table (or whatever you named these tables). Save the form as Customer. Your form should resemble that in Figure 2-3.

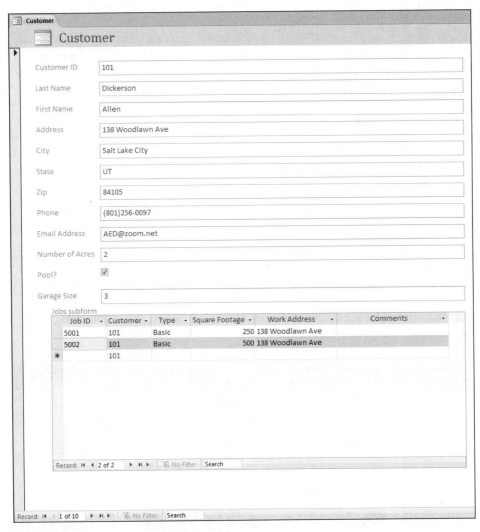

FIGURE 2-3 Customer form and Jobs subform

Query 1

Create a query called Customers with Pool and More than 2 Acres. The query should display fields for the Last Name, First Name, Address, City, and State of all customers who have a pool and more than two acres of property. Your output should resemble that in Figure 2-4, although your data will be different.

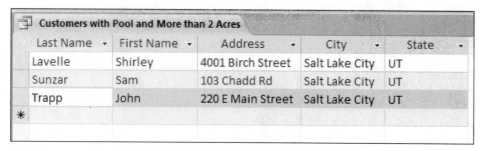

Last Name ▾	First Name ▾	Address ▾	City ▾	State ▾
Lavelle	Shirley	4001 Birch Street	Salt Lake City	UT
Sunzar	Sam	103 Chadd Rd	Salt Lake City	UT
Trapp	John	220 E Main Street	Salt Lake City	UT

FIGURE 2-4 Customers with Pool and More than 2 Acres query

Query 2

Create a query called Number of Jobs. Specify a type of concrete and count the number of jobs for that type. Your query should prompt for the type of concrete so that the owners can run the query for different types. Note the column heading change from the default setting provided by the query generator. If you enter "Basic" as the type of concrete, your output should look like that in Figure 2-5, although your data will be different.

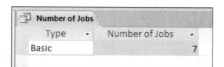

Type ▾	Number of Jobs ▾
Basic	7

FIGURE 2-5 Number of Jobs query

Query 3

Create a query called Most Lucrative Product that specifies each type of concrete and then shows the total square footage and total cost for all jobs that use each type of concrete. (Hint: The Total Cost field is a calculated field with aggregated data.) Note the column heading change from the default setting provided by the query generator. Sort your output to show the most lucrative product first. Your output should resemble the format shown in Figure 2-6, but the data will be different.

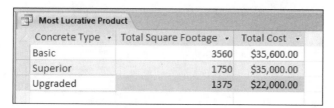

Concrete Type ▾	Total Square Footage ▾	Total Cost ▾
Basic	3560	$35,600.00
Superior	1750	$35,000.00
Upgraded	1375	$22,000.00

FIGURE 2-6 Most Lucrative Product query

Query 4

Create a query that lists the jobs scheduled in a particular month for either of the company's owners. In Figure 2-7, the query is called April Jobs for Candy. The query should display fields for the Job ID, Work

Address, and Start Date for each job in a particular month. Your output should resemble the format shown in Figure 2-7, but the data will be different.

April Jobs for Candy		
Job ID	Work Address	Start Date
5001	138 Woodlawn Ave	4/15/2013
5002	138 Woodlawn Ave	4/17/2013
5003	121 Chaucer Lane	4/20/2013
5004	26 Julie Court	4/22/2013
5005	55 Peachtree Court	4/27/2013

FIGURE 2-7   April Jobs for Candy query

Report

Create a report based on a query that displays the first and last name of customers, the type of concrete they selected for their work, the square footage of the jobs, and the cost of the jobs (a calculated field). Save the query as For Report. Using all the data in the output of the query, create a grouped report. Group the report on Last Name, and bring the First Name to the same grouping line as Last Name. Give the report a title of Customer Payments, and save the report using the same filename. Create a subtotal for each customer's total cost, and create a grand total for all customers at the bottom of the report. (The grand total is not shown in Figure 2-8.) Make sure that all data and column headings are visible, and format your output appropriately, such as including currency signs. Depending on your data, your output should resemble that in Figure 2-8; only a portion of the report is shown here.

Customer Payments

Last Name	First Name	Concrete Type	Square Footage	Cost
Dickerson	Allen			
		Basic	500	$5,000.00
		Basic	250	$2,500.00
				$7,500.00
Faber	Dale			
		Superior	1250	$25,000.00
				$25,000.00
Hearn	Arthur			
		Upgraded	500	$8,000.00
				$8,000.00
Lavelle	Shirley			
		Upgraded	150	$2,400.00
		Basic	100	$1,000.00
		Superior	125	$2,500.00
				$5,900.00
Nelson	Janice			
		Upgraded	300	$4,800.00
		Upgraded	300	$4,800.00
				$9,600.00

FIGURE 2-8   Customer Payments report

ASSIGNMENT 3: MAKING A PRESENTATION

Create a presentation that explains the database to Candy and Carl. Demonstrate how they can use the database by running the queries and generating a report. Discuss future improvements and additions to the database. Your presentation should take fewer than 10 minutes, including a brief question-and-answer period.

DELIVERABLES

Assemble the following deliverables for your instructor, either electronically or in printed form:

1. Word-processed design of tables
2. Tables created in Access
3. Form and subform: Customer (with Jobs subform)
4. Query 1: Customers with Pool and More than 2 Acres
5. Query 2: Number of Jobs
6. Query 3: Most Lucrative Product
7. Query 4: April Jobs for Candy
8. Query 5: For Report
9. Report: Customer Payments
10. Presentation materials
11. Any other required printouts or electronic media

Staple all pages together. Include your name and class number at the top of the page. Make sure that your electronic media are labeled, if required.

CASE **3**

COLLECTIVE COUPONS TRACKING DATABASE

Designing a Relational Database to Create Tables, Forms, Queries, and Reports

PREVIEW

In this case, you will design a relational database for a company that markets and sells coupons to groups on the Internet. After your database design is completed and correct, you will create database tables and populate them with data. You will produce one form with a subform that allows you to book coupon deals. You will also create queries to help the company answer some important questions: Which deals are available for a specified maximum price? How good a bargain is a deal compared with the regular price? Which deals have at least 100 people signed up so the deal can run? What is the most popular deal? Finally, you will create a query and then a report that lists all customers who have signed up for current deals.

PREPARATION

- Before attempting this case, you should have some experience in database design and in using Microsoft Access.
- Complete any part of Tutorial A that your instructor assigns.
- Complete any part of Tutorial B that your instructor assigns, or refer to the tutorial as necessary.
- Refer to Tutorial F as necessary.

BACKGROUND

The Internet is a great avenue for marketing and sales, and e-commerce has exploded since the mid-1990s. Recently, new companies have begun to offer discounted coupons to registered members. These coupon deals originate in local businesses, which offer products and services for about half the normal price if the coupon company can get 100 or more people to sign up for a deal. For example, suppose that a local restaurant wants to increase its patronage during the middle of the week, when the dining room usually is not full. The restaurant will offer the coupon company half-price deals on mid-week dinners, such as a $100 gift certificate for $50, if the coupon company can get at least 100 people to sign up for the deal. By using e-mail and Facebook, coupon companies can often find 100 people who want to take advantage of a particular bargain.

You have been hired as a summer intern to create a database tracking system for an Internet coupon company called Collective Coupons. The owner, Charlene Clayton, saw your resume on your university Web site. She thinks that your experience with database design, implementation, and Microsoft Access makes you a perfect fit for the job. Charlene wants you to design the database first and then implement a number of forms, queries, and a report.

On your first day at work, you interview Charlene to find out what she wants from the database:

YOU: What sort of information do you need to keep about your business?

CHARLENE: We have a spreadsheet of all our customers, including their names and e-mail addresses. Each record is designated with a customer number so we can keep them straight. So far we have over 3,000 customers registered with us, so the business is really taking off well! We don't need additional information about the customers because payment for the deals occurs only after 100 people sign up and the deal is accepted. Those payments are handled by another system we already have in place. But we would like an easy way to record who signs up for each deal.

YOU: A form with a subform would be a good way to do that. What information do you need to keep about the deals you solicit from local businesses?

CHARLENE: We are starting to have so many deals that we need to keep them straight. Some deals have the same name because the business might run the same deal later in the year. We keep a description of the deal, the city where the deal is offered and run, the available dates, and the price of the deal versus the regular price so customers can see what a bargain it is.

YOU: How do you track who signs up for the deals?

CHARLENE: Again, using a spreadsheet, we keep a list of customers who sign up for the deals. But there isn't an easy way to count all the deals, and we need to be able to do that.

YOU: You can easily count all the customers who sign up for the deals by using a query. What other information do you want to get out of this database?

CHARLENE: I'd like to know what deals are available for a certain price. We like to advertise that price; people e-mail us with that question all the time. Also, we want to be able to calculate the savings that customers are getting with a deal. We would advertise these savings heavily since our customers love good bargains. Of course, we need to know which deals are running, so figuring out which deals have at least 100 sign-ups is essential. We'd also like to know which deals are the most popular so we can ask those businesses for more future deals.

YOU: Queries can handle all those questions easily. Do you need any reports?

CHARLENE: We would like to see a report of everyone who has signed up for the current deals.

ASSIGNMENT 1: CREATING THE DATABASE DESIGN

In this assignment, you will design your database tables using a word-processing program. Pay close attention to the logic and structure of the tables. Do not start developing your Access database in Assignment 2 before getting feedback from your instructor on Assignment 1. Keep in mind that you will need to examine the requirements in Assignment 2 to design your fields and tables properly. It is good programming practice to look at the required outputs before beginning your design. When designing the database, observe the following guidelines:

- First, determine the tables you will need by listing the name of each table and the fields it should contain. Avoid data redundancy. Do not create a field if it can be created by a "calculated field" in a query.
- You will need to record the transactions in a separate table. Avoid duplicating data.
- Document your tables using the table feature of your word processor. Your tables should resemble the format shown in Figure 3-1.
- You must mark the appropriate key field(s) by entering an asterisk (*) next to the field name.
- Print the database design if your instructor requires it.

Table Name	
Field Name	Data Type (text, numeric, currency, etc.)
...	...
...	...

FIGURE 3-1 Table design

NOTE

Have your design approved before beginning Assignment 2; otherwise, you may need to redo Assignment 2.

ASSIGNMENT 2: CREATING THE DATABASE, QUERIES, AND REPORT

In this assignment, you will first create database tables in Access and populate them with data. Next, you will create a form, queries, and a report.

Assignment 2A: Creating Tables in Access

In this part of the assignment, you will create your tables in Access. Use the following guidelines:

- Create at least five deals offered by Collective Coupons.
- Create more than 100 customers.
- Make sure that at least 100 customers sign up for some of the deals.
- Appropriately limit the size of the text fields; for example, a customer number does not need the default length of 255 characters.
- Print all tables if your instructor requires it.

Assignment 2B: Creating Forms, Queries, and a Report

You will generate one form with a subform, five queries, and one report, as outlined in the Background section of this case.

Form

Create a form and subform based on your Deals table and Sign-ups table (or whatever you named these tables). Save the main form as Deals. Your form should resemble that in Figure 3-2.

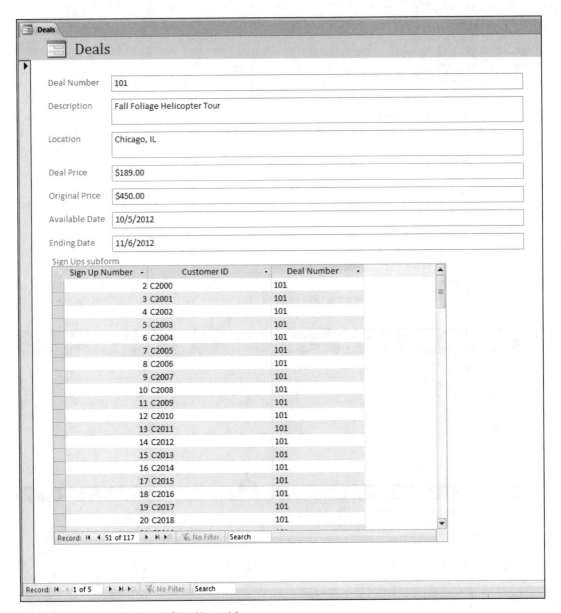

FIGURE 3-2   Deals form and Sign Ups subform

Query 1

Create a query called Maximum Price. This query should prompt the user for a maximum price and then display fields for the Description, Location, Deal Price, Available Date, and Ending Date of all deals under the specified price. For example, if you enter $50 at the prompt, your output should resemble that in Figure 3-3, although your data will be different.

Description	Location	Deal Price	Available Date	Ending Date
18 Hole Golf Heritage Golf Center	Chicago, IL	$28.00	9/2/2012	9/15/2012
Hamburgers, Etc Restaurant	Chicago, IL	$12.00	9/20/2012	9/22/2012
Segway City Tour - 3 hours	Chicago, IL	$35.00	9/18/2012	9/19/2012
Spas Unlimited	Chicago, IL	$25.00	10/1/2012	10/10/2012

FIGURE 3-3 Maximum Price query

Query 2

Create a query called Percentage Bargain. List all the available deals, including their Deal Number, Description, Location, Deal Price, and Original Price, and then calculate a percentage in the Bargain column. The bargain is the percentage difference between the Deal Price and Original Price. Your output should look like that in Figure 3-4, although your data will be different.

Deal Number	Description	Location	Deal Price	Original Price	Bargain
101	Fall Foliage Helicopter Tour	Chicago, IL	$189.00	$450.00	58.00%
102	18 Hole Golf Heritage Golf Center	Chicago, IL	$28.00	$64.00	56.25%
103	Hamburgers, Etc Restaurant	Chicago, IL	$12.00	$25.00	52.00%
104	Segway City Tour - 3 hours	Chicago, IL	$35.00	$70.00	50.00%
105	Spas Unlimited	Chicago, IL	$25.00	$50.00	50.00%

FIGURE 3-4 Percentage Bargain query

Query 3

Create a query called Sign-ups Equal to or Over 100. In this query, you need to determine which deals have at least 100 members signed up. Display columns only for the Deal Number and Description in your output. Your output should resemble the format shown in Figure 3-5, but the data will be different.

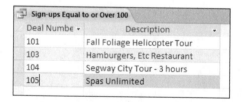

Deal Number	Description
101	Fall Foliage Helicopter Tour
103	Hamburgers, Etc Restaurant
104	Segway City Tour - 3 hours
105	Spas Unlimited

FIGURE 3-5 Sign-ups Equal to or Over 100 query

Query 4

Create a query called Most Popular Deal. List a description of the deal and how many people have signed up for it. Sort the output to list the most popular deals first. Note the column heading change from the default setting provided by the query generator. Your output should resemble the format shown in Figure 3-6, but the data will be different.

Most Popular Deal	
Description	Number Signed Up
Spas Unlimited	129
Fall Foliage Helicopter Tour	117
Hamburgers, Etc Restaurant	114
Segway City Tour - 3 hours	100
18 Hole Golf Heritage Golf Center	68

FIGURE 3-6 Most Popular Deal query

Report

Create a report named Notification Report that is based on tables and the Sign-ups Equal to or Over 100 query you already created. Use the query to output the description of each deal and the e-mail addresses and names of customers signed up for the deal. List only the deals that have 100 or more people signed up. Save the query as For Report. Bring the query into the report generator and group records based on the Description field. Depending on your data, your output should resemble that shown in Figure 3-7; only a portion of the report is shown for space purposes.

Notification Report — Wednesday, November 02, 2011 9:53:09 AM

Description	Email	Name
Fall Foliage Helicopter Tour		
	TUMSJTTR1@gmail.com	J Salo
	ONSCKMAP2@verizon.com	S Jones
	TLZMGOOT2@verizon.com	J Legace
	IEIEWFHW3@comcast.com	P Hasek
	EEXRNLHM2@verizon.com	E Kiprusoff
	IYZAUBON2@verizon.com	W Labarbera
	BIRRRWLW3@comcast.com	O O'Neal
	KQLSJFKI1@gmail.com	A Jagr
	ZGYRRXIC2@verizon.com	L Pierzynski
	XDCMRUYU3@comcast.com	Z Fernandez
	PJTVJWXZ2@verizon.com	U Alfredsson
	BPBDNXWP2@verizon.com	A Joseph
	HQXWVOFD1@gmail.com	U McCabe
	XCAAFLSU3@comcast.com	P Wright
	TOHDCKAY1@gmail.com	T Visnovsky
	CHUTSNZB3@comcast.com	E Eaton
	BFLYXFLT1@gmail.com	M Gagne
	QSXVEVVW1@gmail.com	W Luongo
	HOWXSAQT3@comcast.com	R Lundqvist
	JYGAQQLH2@verizon.com	N Bosh
	MFEHGGLY3@comcast.com	R Duncan

FIGURE 3-7 Notification report

ASSIGNMENT 3: MAKING A PRESENTATION

Create a presentation that explains the database to Charlene and the employees at Collective Coupons. Discuss how you would move this prototype database to the Web. Your presentation should take fewer than 10 minutes, including a brief question-and-answer period.

DELIVERABLES

Assemble the following deliverables for your instructor, either electronically or in printed form:

1. Word-processed design of tables
2. Tables created in Access
3. Form and subform: Deals (with Sign Ups subform)
4. Query 1: Maximum Price
5. Query 2: Percentage Bargain
6. Query 3: Sign-ups Equal to or Over 100
7. Query 4: Most Popular Deal
8. Query 5: For Report
9. Report: Notification
10. Presentation materials
11. Any other required printouts or electronic media

Staple all pages together. Include your name and class number at the top of the page. Make sure that your electronic media are labeled, if required.

THE MOVABLE ART DATABASE

Designing a Relational Database to Create Tables, Forms, Queries, and Reports

PREVIEW

In this case, you will design a relational database for a business that moves pieces of fine art to museums for temporary exhibits. After your database design is completed and correct, you will create database tables and populate them with data. Then you will produce one form with a subform, six queries, and two reports. The queries will answer questions such as which paintings were created by particular artists and which paintings are loaned to particular museums. Other queries will list the paintings shipped in each order and the number of paintings shipped in each order. You will also produce two reports based on queries. One report will list the paintings on loan to various museums, and the other will list the costs of shipping the paintings.

PREPARATION

- Before attempting this case, you should have some experience in database design and in using Microsoft Access.
- Complete any part of Tutorial A that your instructor assigns.
- Complete any part of Tutorial B that your instructor assigns, or refer to the tutorial as necessary.
- Refer to Tutorial F as necessary.

BACKGROUND

Your uncle, Roger, and his business partner, Al, own and run a company that ships fine art. Some paintings are worth millions of dollars; moving precious art requires skill and patience. The company is in demand because more museums are borrowing art to develop temporary exhibits instead of buying art. Because of cutbacks in government grants and arts funding and decreased donations from private sources, museums have to make their budgets stretch. By borrowing art for special exhibits that last a few months and charging extra admission, museums can offer fresh presentations and control their spending.

Roger and Al have run the business for over 20 years, and they still use old-fashioned methods for keeping track of it. Roger saw you at a family reunion last month and heard you talk about your recent class in database design and implementation. He offered you a summer internship to create and implement a database system that will track art shipments.

Here's how the business works: Roger and Al work with a group of wealthy people who own valuable art and who agree to loan their paintings to various museums around the United States. The art patrons want to remain anonymous to the museums, so Roger and Al do not want to record personal information about the patrons in the database. Instead, each patron will be represented by an identification number in the database.

Roger and Al have a list of all the paintings that each patron makes available for loan. The list includes the name of each painting, the artist, and the year the painting was finished. The year of creation is not known for some of the older paintings. Roger and Al work with a number of prominent museums around the country, including the Metropolitan Museum of Art in New York City and a new museum called Crystal Bridges in Bentonville, Arkansas. The men keep a list of the museums they work with, along with addresses, phone numbers, and names of curators.

When a request for art arrives from a museum, Roger and Al set up a shipping log that contains the shipping date and destination. The log also notes whether the museum wants additional liability insurance, which costs $500 per shipment. Each shipment usually includes more than one painting. The business charges a flat rate of $350 to ship each painting.

In addition to having you create the database design and implement the database tables, Roger and Al have specific requirements for the project. First, they want to be able to create shipment orders and specify which paintings go with a shipment. You suggest that a form would be an excellent way to enter the data.

A common problem for Roger and Al is determining which paintings are available for loan by particular artists. Museums often run special exhibits that concentrate on selected artists. You recommend a query to solve the problem. Roger and Al have other common questions: What paintings created before 1900 will be shipped to a museum in a particular state? Which paintings are included in a particular shipment? You are confident that a select query and a parameter query can answer those questions. Roger and Al also need to know how many paintings are shipped in each order.

Finally, your uncle and his partner would like two reports. One report should list museums and the paintings they are scheduled to borrow. Another report will list the amount of money each museum is charged for shipping costs and insurance.

ASSIGNMENT 1: CREATING THE DATABASE DESIGN

In this assignment, you will design your database tables using a word-processing program. Pay close attention to the logic and structure of the tables. Do not start developing your Access code in Assignment 2 before getting feedback from your instructor on Assignment 1. Keep in mind that you will need to examine the requirements in Assignment 2 to design your fields and tables properly. It is good programming practice to look at the required outputs before beginning your design. When designing the database, observe the following guidelines:

- First, determine the tables you will need by listing the name of each table and the fields it should contain. Avoid data redundancy. Do not create a field if it can be created by a calculated field in a query.
- You will need transaction tables. Think about the business events that occur with each shipment. Avoid duplicating data.
- Document your tables using the table feature of your word processor. Your tables should resemble the format shown in Figure 4-1.
- You must mark the appropriate key field(s) by entering an asterisk (*) next to the field name. Keep in mind that some tables might need a compound primary key to uniquely identify a record within a table.
- Print the database design.

Table Name	
Field Name	Data Type (text, numeric, currency, etc.)
...	...
...	...

FIGURE 4-1 Table design

N O T E

Have your design approved before beginning Assignment 2; otherwise, you may need to redo Assignment 2.

ASSIGNMENT 2: CREATING THE DATABASE, QUERIES, AND REPORT

In this assignment, you will first create database tables in Access and populate them with data. Next, you will create a form and subform, six queries, and two reports.

Assignment 2A: Creating Tables in Access

In this part of the assignment, you will create your tables in Access. Use the following guidelines:

- Enter data for at least six museums. Use the Internet to find actual museums and data.
- Use the Internet to find at least 40 paintings from actual artists to populate your database table. Wikipedia is a good starting point as a data source, although you should corroborate your findings by checking other Web sites.

- Each museum should have at least one shipment, and a few should have more than one. Make sure that each shipment has multiple paintings.
- Appropriately limit the size of the text fields; for example, a museum ID number does not need the default length of 255 characters.
- Print all tables if your instructor requires it.

Assignment 2B: Creating Forms, Queries, and Reports

You will generate one form with a subform, six queries, and two reports, as outlined in the Background section of this case.

Form

Create a form and subform based on your Shipping table and Shipping Line Item table (or whatever you named these tables). Save the form as Shipping. Your form should resemble that in Figure 4-2.

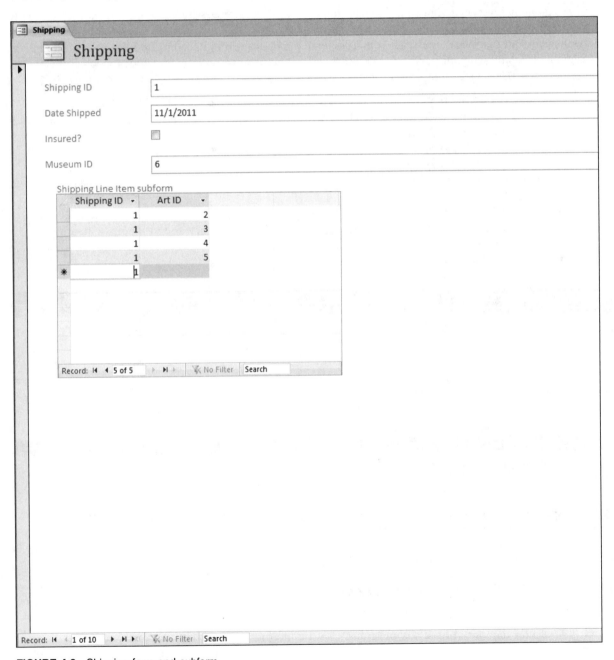

FIGURE 4-2 Shipping form and subform

Query 1

Create a select query that displays a list of paintings, their artists, and the museums that have borrowed the paintings. For example, in Figure 4-3, the query is called Paintings by Picasso and Warhol. Your output should resemble that in Figure 4-3, although the data will be different.

Painting	Artist	Museum Name
Garçon à la pipe	Pablo Picasso	National Gallery
Nude, Green Leaves and Bust	Pablo Picasso	Walters Art Museum
Dora Maar au Chat	Pablo Picasso	Art Institute of Chicago
Eight Elvises	Andy Warhol	Crystal Bridges
Les Noces de Pierrette	Pablo Picasso	Crystal Bridges
Yo, Picasso	Pablo Picasso	Walters Art Museum
Turquoise Marilyn	Andy Warhol	Metropolitan Museum of Art
Green Car Crash (Green Burning Car I)	Andy Warhol	Seattle Art Museum
Au Lapin Agile	Pablo Picasso	Metropolitan Museum of Art
Acrobate et jeune Arlequin[39]	Pablo Picasso	Walters Art Museum
Femme aux Bras Croisés	Pablo Picasso	Seattle Art Museum
Le Rêve[44]	Pablo Picasso	Walters Art Museum
Femme assise dans un jardin	Pablo Picasso	Seattle Art Museum
Men in Her Life	Andy Warhol	National Gallery

FIGURE 4-3   Paintings by Picasso and Warhol query

Query 2

Create a select query called Pre-1900 and NY Paintings that lists art created before 1900 that is currently on loan to New York museums. The query lists the identification number of the painting, the name of the painting, the artist, the year the painting was created, the state where the museum is located, and the museum name. Your output should resemble that in Figure 4-4, although the data will be different.

ID	Painting	Artist	Year	Museum State	Museum Name
6	Bal du moulin de la Galette[11]	Pierre-Auguste Renoir	1876	NY	Metropolitan Museum of Art

FIGURE 4-4   Pre-1900 and NY Paintings query

Query 3

Create a parameter query called Shipping List by ID that includes a list of paintings, their shipment dates, and whether the shipment has been insured. This query should prompt the user to input the shipping ID. Your output should resemble the format shown in Figure 4-5, but the data will be different.

Painting	Date Shipped	Insured?
Yo, Picasso	2/13/2012	☐
Turquoise Marilyn	2/13/2012	☐
Le Bassin aux Nymphéas	2/13/2012	☐
Rideau, Cruchon et Compotier[31]	2/13/2012	☐
Vase with Fifteen Sunflowers	2/13/2012	☐
White Center (Yellow, Pink and Lavender on Rose)	2/13/2012	☐

FIGURE 4-5   Shipping List by ID query

Query 4

Create a query called Number of Paintings Shipped that lists shipping ID numbers and the total number of paintings in each shipment. Note the column heading change from the default setting provided by the query generator. Your output should resemble the format shown in Figure 4-6, but the data will be different.

Shipping ID	Number of Paintings
1	4
2	3
3	6
4	6
5	2
6	4
7	2
8	1
9	1
10	2

FIGURE 4-6   Number of Paintings Shipped query

Report 1

Create a report named Art Exhibits that displays headings for Museum Name and Painting. First, create a query to display the information. Save the query as Art Exhibits, bring the query data into a report, and group the report data by museum name. Make sure that all column headings and data are visible. Depending on the data, your output should resemble that in Figure 4-7.

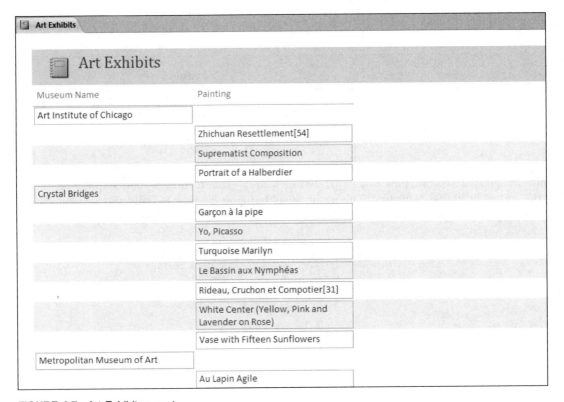

FIGURE 4-7   Art Exhibits report

Report 2

Create a report called Shipping Costs that lists museums, their full address, shipping ID numbers, and total cost of shipping to each museum. First, you need to create a query to calculate the total shipping cost. Each painting costs $350 to ship and $500 extra if the museum purchases insurance to cover shipping. Use an IF statement to add the insurance cost. Bring the query into a report and group by museum name. Bring the address up to the Museum Name level on the report. Depending on how you create the report, the Total Cost column may display logical data. If so, switch to Design view and remove the report's Total Cost control, which is the box in the Detail band. Insert a text box control in place of the deleted control; the source of the new control should be the Total Cost calculated field. Make sure that all column headings and data are visible. Depending on your data, the report should resemble that in Figure 4-8.

Shipping Costs

Museum Name	Museum Address				Shipping ID	Total Cost
Art Institute of Chicago	111 South Michigan Avenue	Chicago	IL	60603		
					2	$1,550.00
Crystal Bridges	600 Museum Way	Bentonville	AR	72712		
					9	$850.00
					4	$2,100.00
Metropolitan Museum of Art	1000 Fifth Avenue	New York	NY	10028		
					7	$700.00

FIGURE 4-8   Shipping Costs report

ASSIGNMENT 3: MAKING A PRESENTATION

Create a presentation that explains the database to Roger and Al. Describe the design of your database tables and include instructions for using the database. Discuss future improvements to the database, such as placing the company's list of available art and a shipping form on the Web. Your presentation should take fewer than 10 minutes, including a brief question-and-answer period.

DELIVERABLES

Assemble the following deliverables for your instructor, either electronically or in printed form:

1. Word-processed design of tables
2. Tables created in Access
3. Form and subform: Shipping
4. Query 1: Paintings by Picasso and Warhol
5. Query 2: Pre-1900 and NY Paintings
6. Query 3: Shipping List by ID
7. Query 4: Number of Paintings Shipped
8. Query 5: Art Exhibits
9. Query 6: Shipping Costs
10. Report 1: Art Exhibits
11. Report 2: Shipping Costs
12. Presentation materials
13. Any other required printouts or electronic media

Staple all the pages together. Include your name and class number at the top of the page. Make sure that your electronic media are labeled, if required.

THE ANIMAL LOAN DATABASE

Designing a Relational Database to Create Tables, Forms, Queries, and Reports

PREVIEW

In this case, you will design a relational database that serves as a tracking system for a national consortium of zoos that lend and borrow animals for temporary exhibits. After your database design is completed and correct, you will create database tables and populate them with data. Then you will produce one form, eight queries, and two reports. The form will allow zoos to request animals for loan. The queries will list monkeys on loan to the zoos, the animals on loan to a specified zoo, the regions around the world from which the loaned animals originate, and the most popular animals on loan. Queries will also report the length of each animal loan and the zoos with the most animal loans. The two reports, both based on queries, will list awards given to volunteers and display the hours worked by volunteers in the current month.

PREPARATION

- Before attempting this case, you should have some experience in database design and in using Microsoft Access.
- Complete any part of Tutorial A that your instructor assigns.
- Complete any part of Tutorial B that your instructor assigns, or refer to the tutorial as necessary.
- Refer to Tutorial F as necessary.

BACKGROUND

Did you know that many of the animals you see in a zoo might not actually belong to that zoo? Zoos from all over the world exchange animals free of charge. Zoos lend animals because they might have too many of a certain species or they might need to move animals before renovating an exhibit. Zoos might borrow animals because they want to draw more visitors by rotating their exhibits. A zoo might even borrow animals for breeding purposes. During the recent financial recession, zoo attendance is at an all-time high because the entry fee is typically much less than that of a theme park. You have been hired by a nonprofit organization called Zoo-to-You to design and implement a database that keeps track of zoo animal exchanges and the volunteers who work with the animals.

You meet with Zoo-to-You CEO Lauren Magee to learn more about their operation. Lauren explains that a number of zoos participate in animal exchanges. Each zoo first registers the animals it has available. The registration list includes each animal, its region of origin, and its species. You can assume that participating zoos have an abundant supply of animals to loan, so they don't need to keep track of the number of animals "in stock."

Zoos that want to borrow animals call Zoo-to-You with their requests. Lauren's employees record information about each requesting zoo, the animals it wants to borrow, and the dates that each animal would be borrowed. The names and addresses of participating zoos are kept in the office in a Rolodex file.

Volunteers play an important role in making sure that all animals are transported safely and cared for in their temporary homes. Lauren's company keeps a list of all volunteers, along with their addresses, telephone numbers, e-mail addresses, and their starting date as volunteers. Zoo-to-You uses a binder to record the work hours of each volunteer.

Lauren would like to have a number of inputs and outputs from the database after you have designed and populated the tables. First, she would like an easy way to enter requests that arrive for animal loans. You tell her that she can easily enter requests using a form and subform.

A number of questions need to be answered with queries. First, Lauren would like a list of all monkeys on loan to various zoos. Monkeys are a favorite draw for zoo visitors, and Lauren would like to use the list to advertise for donations to Zoo-to-You. Next, she wants to be able to find out quickly which animals are on loan at a specific zoo. Ideally, the query should prompt the user to enter a zoo name and return a complete list of animals supplied by Zoo-to-You.

Lauren and her workers are curious to know the regions around the world from which most loaned animals originate and which animals are the most popular to loan. Queries can answer both questions. Also, Lauren would like to see which zoo has the most animals on loan; she suspects it is the world-famous San Diego Zoo.

The volunteer coordinator, Joan, would like to award gold and silver pens to volunteers depending on their length of service with Zoo-to-You. Joan would like a report that lists the volunteers and the type of pen they should receive. In addition, Joan would like a list of the dates that each volunteer has worked, including the number of hours worked per day and the total hours worked.

Lauren and the team at Zoo-to-You are confident that you will be able to assist in creating the database. Good luck!

ASSIGNMENT 1: CREATING THE DATABASE DESIGN

In this assignment, you will design your database tables using a word-processing program. Pay close attention to the logic and structure of the tables. Do not start developing your Access code in Assignment 2 before getting feedback from your instructor on Assignment 1. Keep in mind that you will need to examine the requirements in Assignment 2 to design your fields and tables properly. It is good programming practice to look at the required outputs before beginning your design. When designing the database, observe the following guidelines:

- First, determine the tables you will need by listing the name of each table and the fields it should contain. Avoid data redundancy. Do not create a field if it can be created by a calculated field in a query.
- You must create a few transaction tables to cover each aspect of Zoo-to-You described in the Background section. Avoid duplicating data.
- Document your tables using the table feature of your word processor. Your tables should resemble the format shown in Figure 5-1.
- You must mark the appropriate key field(s) by entering an asterisk (*) next to the field name. Keep in mind that some tables might need a compound primary key to uniquely identify a record within a table.
- Print the database design.

Table Name	
Field Name	Data Type (text, numeric, currency, etc.)
...	...
...	...

FIGURE 5-1   Table design

NOTE

Have your design approved before beginning Assignment 2; otherwise, you may need to redo Assignment 2.

ASSIGNMENT 2: CREATING THE DATABASE, QUERIES, AND REPORTS

In this assignment, you will first create database tables in Access and populate them with data. Next, you will create a form with a subform, eight queries, and two reports.

Assignment 2A: Creating Tables in Access

In this part of the assignment, you will create your tables in Access. Use the following guidelines:

- Choose one type of animal, such as a mammal or reptile, and create at least 50 records for that type of animal. Consider visiting a zoo's Web site to find animals kept at that zoo.

- Choose at least 10 different zoos by researching them on the Internet. Enter the proper name of each zoo along with their city and state.
- Create records for a large number of animals that can be exchanged among the 10 zoos. You might decide to use the =randbetween function in Microsoft Excel to generate random numbers within a range you specify. You then can import the Excel-generated data into a new Access table. (For details, see the section on importing data in Tutorial B.)
- Create records for at least 10 volunteers and include their work schedules for a particular month. Assume that the year has recently begun and that the work schedules cover the month of January. Record at least seven different dates and times that some of the volunteers worked near the beginning of the month. Make some volunteers work more than one day.
- Appropriately limit the size of the text fields; for example, a zoo ID number does not need the default length of 255 characters.
- Print all tables if your instructor requires it.

Assignment 2B: Creating Forms, Queries, and Reports

You will create a form with a subform, eight queries, and two reports, as outlined in the Background section of this case.

Form

Create a form and subform based on your tables that registers participating zoos and animal loans. Use all the fields in your tables to create the form and subform. Name the form Zoos. Your data will vary, but the output should resemble that in Figure 5-2.

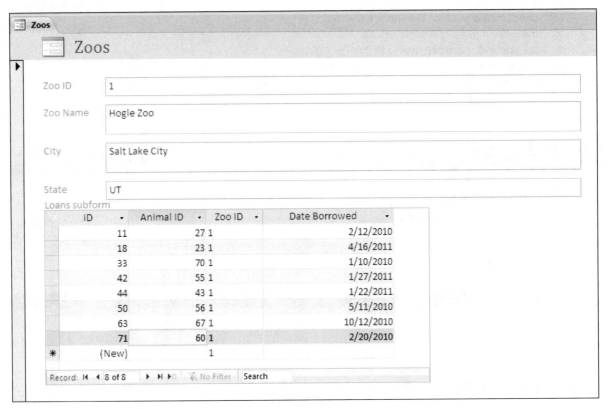

FIGURE 5-2 Zoos form

Query 1

Create a query called Monkeys on Loan that displays headings for Animal, Zoo Name, City, and State. Consider using wildcards for this query. Your data will differ, but the output should resemble that in Figure 5-3.

Animal	Zoo Name	City	State
Common Squirrel Monkey	Bronx Zoo	Bronx	NY
Bolivian Gray Titi Monkey	Great Plains Zoo	Sioux Falls	SD
Spider Monkey	Pittsburgh Zoo	Pittsburgh	PA
Spider Monkey	Pittsburgh Zoo	Pittsburgh	PA
Bolivian Gray Titi Monkey	Zoo New England	Boston	MA
Eastern Black and White Colobus Monkey	Detroit Zoo	Detroit	MI
Brown Tufted Capuchin Monkey	Pittsburgh Zoo	Pittsburgh	PA

FIGURE 5-3 Monkeys on Loan query

Query 2

Create a query called Animals on Loan by Specific Zoo. This query should prompt the user to enter a zoo name. Display headings for the zoo name, the animal, and the region from which the animal originates. Your data will differ, but the output should resemble that in Figure 5-4.

Zoo Name	Animal	Region
Hogle Zoo	Desert Cottontail	North America
Hogle Zoo	Cougar	North America
Hogle Zoo	Wied's Marmoset	South America
Hogle Zoo	Schmidt's Guenon	Africa
Hogle Zoo	North American Opossum	North America
Hogle Zoo	Short-tailed Bat	South America
Hogle Zoo	White Rhinoceros	Africa
Hogle Zoo	Southern Flying Squirrel	North America

FIGURE 5-4 Animals on Loan by Specific Zoo query

Query 3

Create a query called Most Popular Region that displays headings for regions of the world and the number of animals on loan from that region. Be sure to sort the output to display the most popular region at the top of the list. Your data will differ, but the output should resemble that in Figure 5-5. Note the column heading change from the default setting provided by the query generator.

Region	Number of Animals on Loan
Africa	32
South America	30
North America	19
Europe & Asia	11
Australasia	6
Worldwide	2

FIGURE 5-5 Most Popular Region query

Query 4

Create a query called Most Popular Animal that displays headings for Animal and Number of Animals on Loan. Be sure to sort the output to show the most popular loaned animal at the top of the list. Your data will differ, but the output should resemble that in Figure 5-6. Note the column heading change from the default setting provided by the query generator.

Most Popular Animal	
Animal	Number of Animals on Loan
Cuvier's Gazelle	5
Schmidt's Guenon	5
Black-and-White Ruffed Lemur	4
Bornean Orangutan	4
Straw-Colored Fruit Bat	4
Short-tailed Bat	4
North American Opossum	3
Yellow-bellied Marmot	3
Ring-tailed Lemur	3
Desert Cottontail	3
Wied's Marmoset	3
Mara (Patagonian Cavy)	2
Norway Rat	2
Cacomistle	2
Brown-nosed Coati	2
African Elephant	2
Long-tailed Chinchilla	2
Bolivian Gray Titi Monkey	2
American Bison	2
Navajo Sheep	2
Grey Wolf	2
White-handed Gibbon	2
Southern Three-banded Armadillo	2
Spider Monkey	2

FIGURE 5-6   Most Popular Animal query

Query 5

Create a query called Length of Loans. Display headings for the zoo name, animal, and the number of days on loan. The last field is a calculated field that counts the number of days on loan from the original loan date until today. Sort the output to show the longest loans at the top of the list. Your data will differ, but the output should resemble that in Figure 5-7; only a portion of the data appears in the figure.

Length of Loans		
Zoo Name	Animal	Days on Loan
San Diego Zoo	Thomson's Gazelle	705
San Diego Zoo	Cacomistle	703
Hogle Zoo	Wied's Marmoset	668
Zoo New England	Bolivian Gray Titi Monkey	652
Bronx Zoo	Long-tailed Chinchilla	650
Detroit Zoo	Cuvier's Gazelle	647
Topeka Zoo	Amur (Siberian) Tiger	645
Detroit Zoo	Mara (Patagonian Cavy)	645
Topeka Zoo	Norway Rat	643
Detroit Zoo	Black-and-White Ruffed Lemur	642
Great Plains Zoo	Bolivian Gray Titi Monkey	637
Hogle Zoo	Desert Cottontail	635
Hogle Zoo	Southern Flying Squirrel	627
San Diego Zoo	Black-footed Ferret	619
Zoo New England	American Bison	619
Bronx Zoo	Schmidt's Guenon	616
San Diego Zoo	Brown-nosed Coati	606
National Zoo	Snow Leopard	602
Detroit Zoo	Ring-tailed Lemur	598
Great Plains Zoo	Wied's Marmoset	589
Great Plains Zoo	Rock Hyrax	586
Zoo New England	Kinkajou	583
Topeka Zoo	Chacoan Peccary	581
Bronx Zoo	Common Squirrel Monkey	575

FIGURE 5-7   Length of Loans query

Query 6

Create a query called Zoo with Most Loans. Display headings for Zoo Name and Number of Animals on Loan. Sort the output to show the zoo with the most animals on loan first. Your data will differ, but the output should resemble that in Figure 5-8.

Zoo with Most Loans	
Zoo Name	Number of Animals on Loan
San Diego Zoo	14
Great Plains Zoo	12
Detroit Zoo	11
Topeka Zoo	10
Pittsburgh Zoo	10
National Zoo	10
Bronx Zoo	10
Seneca Park Zoo	9
Hogle Zoo	8
Zoo New England	6

FIGURE 5-8   Zoo with Most Loans query

Report 1

Create a report called Volunteer Awards. First create a query that shows the last and first names of volunteers, their length of service, and their award. The Length as a Volunteer field is a calculated field in the query, and the Award field requires an IF statement. Use the following rule: If volunteers have more than seven years of service, they receive a gold pen. Otherwise, they get a silver pen. Save the query as Volunteer Awards, and then bring the query into a report with the same title. Your data will differ, but the output should resemble that in Figure 5-9.

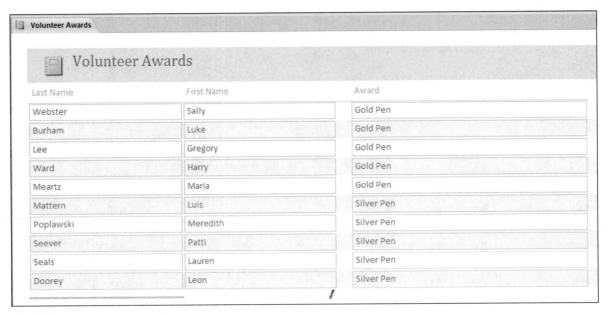

FIGURE 5-9   Volunteer Awards report

Report 2

Create a second volunteer report that is based on a query. The query should list the last and first names of volunteers, the date and starting time for each day they volunteered, and the total hours worked each day. The Total Time field is a calculated field. Save the query as Volunteer January Hours. Group the report on Last Name, and then adjust the output so that the First Name is on the same line as the Last Name. Sum the total time worked by each volunteer. Your data will differ, but the output should resemble that in Figure 5-10. Give the report a title of Volunteer January Hours and save the report with the same name. All data and headings should be visible, as shown in Figure 5-10.

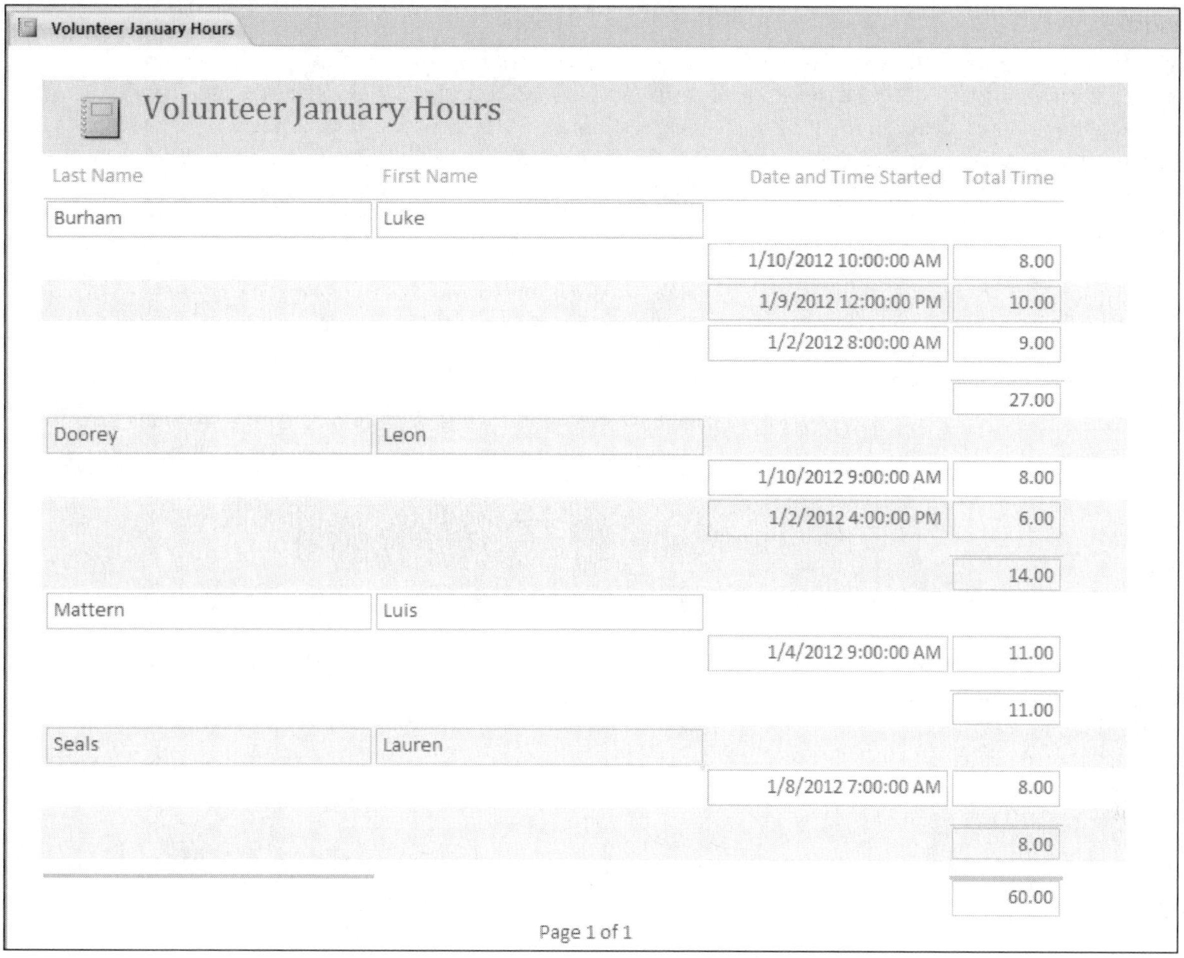
Volunteer January Hours

Volunteer January Hours

Last Name	First Name	Date and Time Started	Total Time
Burham	Luke		
		1/10/2012 10:00:00 AM	8.00
		1/9/2012 12:00:00 PM	10.00
		1/2/2012 8:00:00 AM	9.00
			27.00
Doorey	Leon		
		1/10/2012 9:00:00 AM	8.00
		1/2/2012 4:00:00 PM	6.00
			14.00
Mattern	Luis		
		1/4/2012 9:00:00 AM	11.00
			11.00
Seals	Lauren		
		1/8/2012 7:00:00 AM	8.00
			8.00
			60.00

Page 1 of 1

FIGURE 5-10 Volunteer January Hours report

ASSIGNMENT 3: MAKING A PRESENTATION

Create a presentation for Zoo-to-You. Consider discussing how your system could be a prototype for a Web-based database in which zoos request and manage their own animal exchanges. Your presentation should take fewer than 15 minutes, including a brief question-and-answer period.

DELIVERABLES

Assemble the following deliverables for your instructor, either electronically or in printed form:

1. Word-processed design of tables
2. Tables created in Access
3. Form and subform: Zoos
4. Query 1: Monkeys on Loan
5. Query 2: Animals on Loan by Specific Zoo
6. Query 3: Most Popular Region
7. Query 4: Most Popular Animal
8. Query 5: Length of Loans
9. Query 6: Zoo with Most Loans
10. Query 7: Volunteer Awards
11. Query 8: Volunteer January Hours
12. Report 1: Volunteer Awards
13. Report 2: Volunteer January Hours

Staple all the pages together. Include your name and class number at the top of the page. Make sure that your electronic media are labeled, if required.

PART 2

DECISION SUPPORT CASES USING EXCEL SCENARIO MANAGER

BUILDING A DECISION SUPPORT SYSTEM IN EXCEL

Decision Support Systems (DSS) are computer programs used to help managers solve complex business problems. DSS programs are commonly found in large, integrated packages called enterprise resource planning software that provide information services to an organization. Software packages such as SAP™, Microsoft Dynamics™, and PeopleSoft™ offer sophisticated DSS capabilities. However, many business problems can be modeled for solutions using less complex tools, such as Visual Basic, Microsoft Access, and Microsoft Excel.

A DSS program is actually a model representing a quantitative business problem. The problem can range from finding a desired product mix to sales forecasts to risk analysis, but almost all of the problems examine *financial outcomes*. The model itself contains the data and the algorithms (mathematical processes) needed to solve the problem.

In a DSS program, the user manually inputs data or the program accesses data from a file in the system. The program runs the data through its algorithms and displays output formatted as information; the manager uses this data to decide what action to take to solve the problem. Some sophisticated DSS programs display multiple solutions and recommend one based on predefined parameters.

Managers often find the Excel spreadsheet program particularly useful for their DSS needs. Excel contains hundreds of built-in arithmetic, statistical, logical, and financial functions. It can import data in numerous formats from large database programs, and it can be set up to display well-organized, visually appealing tables and graphs from the output.

This tutorial is organized into four sections:

1. **Spreadsheet and DSS Basics**—This section lets you "get your feet wet" by creating a DSS program in Excel. The program is a cash flow model for a small business looking to expand. You will get an introduction to spreadsheet design, building a DSS, and using financial functions.

2. **Scenario Manager**—Here you will learn how to use the Excel Scenario Manager. A DSS typically gives you one set of answers based on one set of inputs—the real value of the tool lies in its ability to play "what if" and take a comparative look at all the solutions based on all combinations of the inputs. Rather than inputting and running the DSS several times manually, you can use Scenario Manager to run and display the outputs from all possible combinations of the inputs. The output is summarized on a separate worksheet in the Excel workbook.

3. **Practice Using Scenario Manager**—Next, you will be given a new problem to model as a DSS, using Scenario Manager to display your solutions.

4. **Review of Excel Basics**—This section reviews additional information that will help you complete the spreadsheet cases that follow this tutorial. You will learn some basic operations, logical functions, and cash flow calculations.

SPREADSHEET AND DSS BASICS

You are the owner of a thrift shop that resells clothing and housewares in a university town. Many of your customers are college students. Your business is unusual in that sales actually increase during an economic recession. Your cost of obtaining used items basically follows the consumer price index. It is the end of 2012, and business has been very good due to the continuing recession. You are thinking of expanding your business to an adjacent storefront that is for sale, but you will have to apply for a business loan to finance the purchase. The bank requires a projection of your profit and cash flows for the next two years before it will loan you the money to expand, so you have to determine your net income (profit) and cash flows for 2013 and 2014. You decide that your forecast should be based on four factors: your 2012 sales dollars, your cost of goods sold per sales dollar, your estimates of the underlying economy, and the business loan payment amount and interest rate.

Because you will present this model to your prospective lenders, you decide to use an Income and Cash Flow Statements framework. You will input values for two possible states of the economy for 2013 and 2014: R for a continuing recession and B for a "boom" (recovery). Your sales in the recession were growing at 20% per year. If the recession continues and you expand the business, you expect sales to continue growing at 30% per year. However, if the economy recovers, some of your customers will switch to buying "new," so you expect sales growth for your thrift shop to be 15% above the previous year (only 5% growth plus 10% for the business expansion). If you do not expand, your recession or boom growth percentages will only be 20% and 5%, respectively. To determine the cost of goods sold for purchasing your merchandise, which is currently 70% of your sales, you will input values for two possible consumer price outlooks: H for high inflation (1.06 multiplied by the average cost of goods sold) and L for low inflation (1.02 multiplied by the cost of goods sold).

You currently own half the storefront and will need to borrow $100,000 to buy and renovate the other half. The bank has indicated that, depending on your forecast, it may be willing to loan you the money for your expansion at 5% interest during the current recession with a 10-year repayment compounded annually ("R"). However, if the prime rate drops at the start of 2013 because of an economic turnaround ("B"), the bank can drop your interest rate to 4% with the same repayment terms.

As an entrepreneur, an item of immediate interest is your cash flow position with the additional burden of a loan payment. After all, one of your main objectives is to make a profit (Net Income After Taxes). You can use the DSS model to determine if it is more profitable *not* to expand the business.

Organization of the DSS Model

A well-organized spreadsheet will make the design of your DSS model easier. Your spreadsheet should have the following sections:

- Constants
- Inputs
- Summary of Key Results
- Calculations (with separate calculations for Expansion vs. No Expansion)
- Income and Cash Flow Statements (with separate statements for Expansion vs. No Expansion)

Figures C-1 and C-2 illustrate the spreadsheet setup for the DSS model you want to build.

	A	B	C	D
1	**Tutorial Exercise--Collegetown Thrift Shop**			
2				
3	**Constants**	2012	2013	2014
4	Tax Rate	NA	33%	35%
5	Loan Amount for Store Expansion	NA	$100,000	NA
6				
7	**Inputs**	2012	2013	2014
8	Economic Outlook (R=Recession, B=Boom)	NA		NA
9	Inflation Outlook (H=High, L=Low)	NA		NA
10				
11	**Summary of Key Results**	2012	2013	2014
12	Net Income after Taxes (Expansion)	NA		
13	End-of-year Cash on Hand (Expansion)	NA		
14	Net Income after Taxes (No Expansion)	NA		
15	End-of-year Cash on Hand (No Expansion)	NA		
16				
17	**Calculations (Expansion)**	2012	2013	2014
18	Total Sales Dollars	$350,000		
19	Cost of Goods Sold	$245,000		
20	Cost of Goods Sold (as a percent of Sales)	70%		
21	Interest Rate for Business Loan		NA	NA
22				
23	**Calculations (No Expansion)**	2012	2013	2014
24	Total Sales Dollars	$350,000		
25	Cost of Goods Sold	$245,000		
26	Cost of Goods Sold (as a percent of Sales)	70%		

FIGURE C-1   Tutorial skeleton 1

	A	B	C	D
28	**Income and Cash Flow Statements (Expansion)**	**2012**	**2013**	**2014**
29	Beginning-of-year Cash on Hand	NA		
30	Sales (Revenue)	NA		
31	Cost of Goods Sold	NA		
32	*Business Loan Payment*	NA		
33	Income before Taxes	NA		
34	Income Tax Expense	NA		
35	Net Income after Taxes	NA		
36	End-of-year Cash on Hand	$15,000		
37				
38	**Income and Cash Flow Statements (No Expansion)**	**2012**	**2013**	**2014**
39	Beginning-of-year Cash on Hand	NA		
40	Sales (Revenue)	NA		
41	Cost of Goods Sold	NA		
42	Income before Taxes	NA		
43	Income Tax Expense	NA		
44	Net Income after Taxes	NA		
45	End-of-year Cash on Hand	$15,000		

FIGURE C-2   Tutorial skeleton 2

Each spreadsheet section is discussed in detail next.

The Constants Section

This section holds values that are needed for the spreadsheet calculations. These values are usually given to you, and generally do not change for the exercise. However, you can change these values later if necessary; for example, you might need to borrow more or less money for your business expansion (cell C5). For this tutorial, the constants are the Tax Rate and the Loan Amount.

The Inputs Section

The Inputs section in Figure C-1 provides a place to designate the two possible economic outlooks and the two possible inflation outlooks. If you wanted to make these outlooks change by business year, you could leave blanks under both business years. However, as you will see later when you use Scenario Manager, this approach would greatly increase the complexity of interpreting the results. For simplicity's sake, assume that the same outlooks will apply to both 2013 and 2014.

The Summary of Key Results Section

This section summarizes the Year 2 and 3 Net Income after Taxes (profit) and the End-of-year Cash on Hand both for expanding the business and for not expanding. These cells are copied from the Income and Cash Flow Statements section at the bottom of the sheet. Summary sections are frequently placed near the top of a spreadsheet to allow managers to see a quick "bottom line" summary without having to scroll down the spreadsheet to see the final result. Summary sections can also make it easier to select cells for charting.

The Calculations Sections (Expansion and No Expansion)

The following areas are used to compute the following necessary results:

- The Total Sales Dollars, which is a function of the Year 2012 value and the Economic Outlook input
- The Cost of Goods Sold, which is the Total Sales Dollars multiplied by the Cost of Goods Sold (as a percent of Sales)
- The Cost of Goods Sold (as a percent of Sales), which is a function of the Year 2012 value and the Inflation Outlook input
- In addition, the Calculations section for the expansion includes the interest rate, which is also a function of the Economic Outlook input. This interest rate will be used to determine the Business Loan Payment in the Income and Cash Flow Statements section.

You could make these formulas part of the Income and Cash Flow Statements section. However, it makes more sense to use the approach shown here because it makes the formulas in the Income and Cash Flow

Statements less complicated. In addition, when you create other DSS models that include unit costing and pricing calculations, you can enter the formulas in this section to facilitate managerial accounting cost analysis.

The Income and Cash Flow Statements Sections (Expansion and No Expansion)

These sections are the financial or accounting "body" of the spreadsheet. They contain the following values:

- Beginning-of-year Cash on Hand, which equals the *prior* year's End-of-year Cash on Hand.
- Sales (Revenue), which in this tutorial is simply the results of the Total Sales Dollars copied from the Calculations section.
- Cost of Goods Sold, which also is copied from the Calculations section.
- Business Loan Payment, which is calculated using the PMT (Payment) function and the inputs for loan amount and interest rate from the Constants and Calculations sections. Note that only the Income and Cash Flow Statement for Expansion includes a value for Business Loan Payment. If you do not expand, you do not need to borrow the money.
- Income before Taxes, which is Sales minus the Cost of Goods Sold; for the expansion scenarios, you also subtract the Business Loan Payment.
- Income Tax Expense, which is zero when there is no income or the income is negative; otherwise, this value is the Income before Taxes multiplied by the Tax Rate from the Constants section.
- Net Income after Taxes, which is Income before Taxes minus Income Tax Expense.
- End-of-year Cash on Hand, which is Beginning-of-year Cash on Hand plus Net Income after Taxes.

Note that this Income and Cash Flow Statement is greatly simplified. It does not address the issues of changes in Inventories, Accounts Payable, and Accounts Receivable, nor any period expenses such as Selling and General Administrative expenses, utilities, salaries, real estate taxes, insurance, or depreciation.

Construction of the Spreadsheet Model

Next, you will work through three steps to build the spreadsheet model:

1. Make a skeleton or "shell" of the spreadsheet. Save it with a name you can easily recognize, such as TUTC.xlsx or Tutorial C *YourName*.xlsx. When submitting electronic work to an instructor or supervisor, include your last name and first initial in the filename.
2. Fill in the "easy" cell formulas.
3. Then enter the "hard" spreadsheet formulas.

Making a Skeleton or "Shell"

The first step is to set up the skeleton worksheet. The skeleton should have headings, text labels, and constants. Do not enter any formulas yet.

Before you start entering data, you should first try to visualize a sensible structure for your worksheet. In Figures C-1 and C-2, the seven sections are arranged vertically down the page; the item descriptions are in the first column (A), and the time periods (years) are in the next three columns (B, C, and D). This is a widely accepted business practice, and is commonly called a "horizontal analysis." It is used to visually compare financial data side by side through successive time periods.

Because your key results depend on the Income and Cash Flow Statements, you usually set up that section first, and then work upward to the top of the sheet. In other words, you set up the Income and Cash Flow Statements section, then the Calculations section, and then the Summary of Key Results, Inputs, and Constants sections. Some might argue that the Income and Cash Flow Statements should be at the top of the sheet, but when you want to change values in the Constants or Inputs section or examine the Summary of Key Results, it does not make sense to have to scroll to the bottom of the worksheet. When you run the model, you do not enter anything in the Income and Cash Flow Statements—they are all calculations. So, it makes sense to put them last.

Here are some other general guidelines for designing effective DSS spreadsheets:

- Decide which items belong in the Calculations section. A good rule of thumb is that if your items have formulas but do not belong in the Income and Cash Flow Statements, put them in the

Calculations section. Good examples are intermediate calculations such as unit volumes, costs and prices, markups, or changing interest rates.

- The Summary of Key Results section should be just that—*key* results. These outputs help you make good business decisions. Key results frequently include net income before taxes (profit) and end-of-year cash on hand (how much cash your business has). However, if you are creating a DSS model on alternative capital projects, your key results can also include cost savings, net present value of a project, or rate of return for an investment.
- The Constants section holds known values needed to perform other calculations. You use a Constants section rather than just including the values in formulas so that you can input new values if they change. This approach makes your DSS model more flexible.

AT THE KEYBOARD

Enter the Excel skeleton shown in Figures C-1 and C-2.

NOTE

When you see NA (Not Applicable) in a cell, do not enter any values or formulas in the cell. The cells that contain values in the 2012 column are used by other cells for calculations. In this example, you are mainly interested in what happens in 2013 and 2014. The rest of the cells are "Not Applicable."

Filling in the "Easy" Formulas

The next step in building a spreadsheet is to fill in the "easy" formulas. To begin, format all the cells that will contain monetary values as Currency with zero decimal places:

- Constants—C5
- Summary of Key Results—C12 to C15, D12 to D15
- Calculations (Expansion)—C18, C19, D18, D19
- Calculations (No Expansion)—C24, C25, D24, D25
- Income and Cash Flow Statements (Expansion)—B36, C29 to C36, D29 to D36
- Income and Cash Flow Statements (No Expansion)—B45, C39 to C45, D39 to D45

NOTE

With the insertion point in cell C12 (where the $0 appears), note the editing window—the white space at the top of the spreadsheet to the right of the f_x symbol. The cell's contents, whether it is a formula or value, should appear in the editing window. In this case, the window shows =C35.

The Summary of Key Results section (see Figure C-3) will contain the values you calculate in the Income and Cash Flow Statements sections. To copy the cell contents for this section, move your cursor to cell C12, click the cell, type =C35, and press Enter. If you formatted your money cells properly, a $0 should appear in cell C12.

	A	B	C	D
	C12 ▾ f_x =C35			
11	**Summary of Key Results**	**2012**	**2013**	**2014**
12	Net Income after Taxes (Expansion)	NA	$0	
13	End-of-year Cash on Hand (Expansion)	NA		
14	Net Income after Taxes (No Expansion)	NA		
15	End-of-year Cash on Hand (No Expansion)	NA		

FIGURE C-3   Value from cell C35 (Net Income after Taxes) copied to cell C12

Because cell C35 does not contain a value yet, Excel assumes that the empty cell has a numerical value of 0. When you put a formula in cell C35 later, cell C12 will echo the resulting answer. Because Net Income after Taxes (Expansion) for 2014 (cell D35) and its corresponding cell in Summary of Key Results (cell D12) are both directly to the right of the values for 2013, you can either type =D35 into cell D12 or copy cell C12 to D12. To perform the copy operation:

1. Click in a cell or click and drag to select the range of cells you want to copy.
2. Hold down the Control key and press C (Ctrl+C).
3. A moving dashed box called a *marquee* should now be animated over the cell(s) selected for copying.
4. Select the cell(s) where you want to copy the data.
5. Hold down the Control key and press V (Ctrl+V). Cell D12 should now contain $0, but actually it has a reference to cell D35. Click cell D12 and look again at the editing window; it should display =D35.

Cells C14, C15, D14, and D15 represent Net Income after Taxes and End-of-year Cash on Hand for both years of No Expansion; these cells are mirrors of cells C44, C45, D44, and D45 in the last section. Select cell C14, type =C44, and press Enter. Select cell C14 again, use the Copy command, and paste the contents into cell D14 (see Figure C-4).

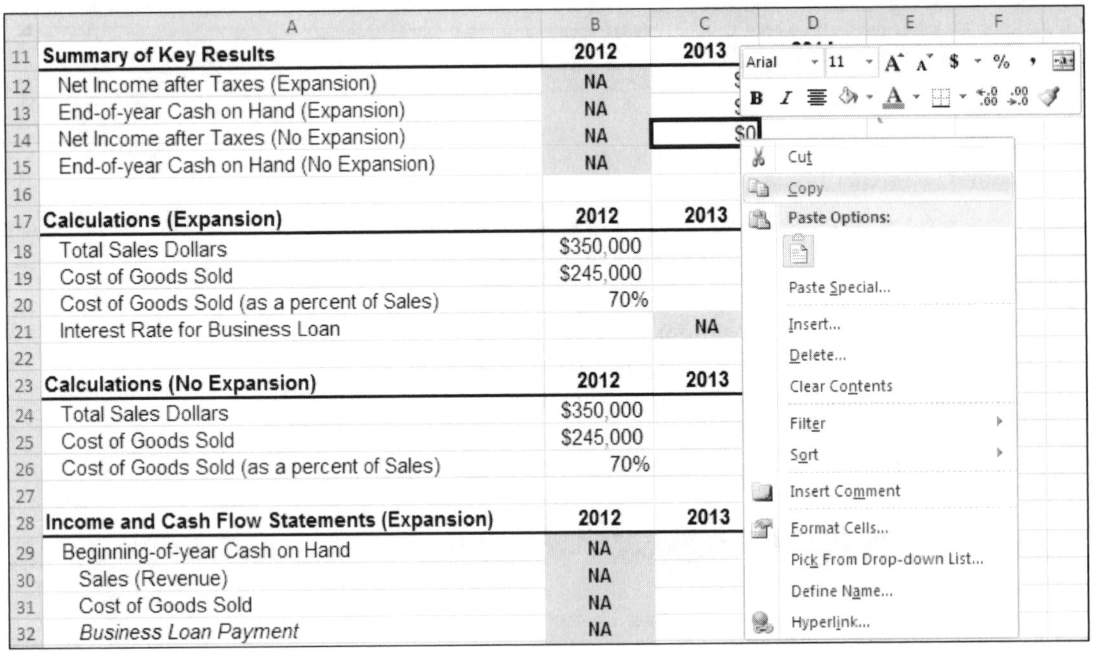

FIGURE C-4   Copying the formula from cell C14 to cell D14

Because Excel uses *relative* cell references by default, copying cell C14 into cell D14 will copy and paste the contents of cell D44 (the cell adjacent to C44) into cell D14. See Figure C-5.

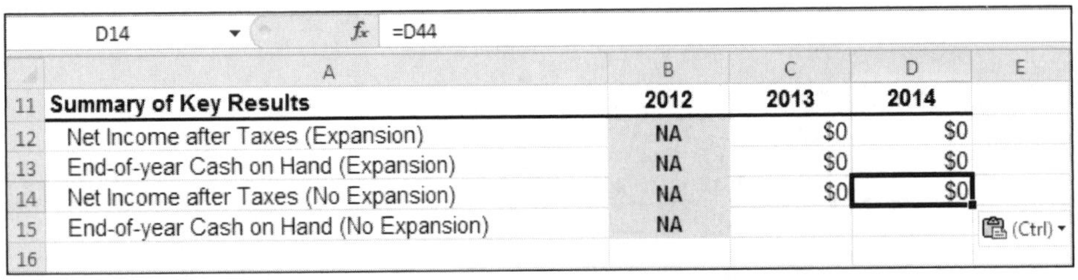

FIGURE C-5   Formula from cell D44 pasted into cell D14

Use the Copy command again, this time downward from cells C14 and D14, to complete cells C15 and D15. If you are successful, the formula in the editing window for cell C15 will be "=C45" and for cell D15 will display "=D45."

You will create the formulas for the two Calculations sections last because they are the hardest formulas. Next, you will create the formulas for the two Income and Cash Flow Statements sections; all the cells in these two sections should be formatted as Currency with zero decimal places.

As shown in Figure C-6, the Beginning-of-year Cash on Hand for 2013 is the End-of-year Cash on Hand for 2012. In cell C29, type =B36. A handy shortcut is to type the "=" sign, immediately move your mouse pointer to the cell you want to designate, and then click the left mouse button. Excel will enter the cell location into the formula for you. This shortcut is especially useful if you want to avoid making a typing error.

	SUM ▾ ✕ ✓ *fx* =B36			
	A	B	C	D
28	**Income and Cash Flow Statements (Expansion)**	**2012**	**2013**	**2014**
29	Beginning-of-year Cash on Hand	NA	=B36	
30	Sales (Revenue)	NA		
31	Cost of Goods Sold	NA		
32	*Business Loan Payment*	NA		
33	Income before Taxes	NA		
34	Income Tax Expense	NA		
35	Net Income after Taxes	NA		
36	End-of-year Cash on Hand	$15,000		
37				
38	**Income and Cash Flow Statements (No Expansion)**	**2012**	**2013**	**2014**
39	Beginning-of-year Cash on Hand	NA		
40	Sales (Revenue)	NA		
41	Cost of Goods Sold	NA		
42	Income before Taxes	NA		
43	Income Tax Expense	NA		
44	Net Income after Taxes	NA		
45	End-of-year Cash on Hand	$15,000		

FIGURE C-6   End-of-year Cash on Hand for 2012 copied to Beginning-of-year Cash on Hand for 2013

Likewise, copy the other three End-of-year Cash on Hand cells to the Beginning-of-year Cash on Hand cells for both Income and Cash Flow Statements (cells D29, C39, and D39).

The Sales (Revenue) cells C30, D30, C40, and D40 are simply copies of cells C18, D18, C24, and D24, respectively, from the Calculations sections (both Expansion and No Expansion). Use the shortcut method to copy these cells. Note that all four cells will display $0 until you enter the formulas in the Calculations sections (see Figure C-7).

	D40 ▾ *fx* =D24			
	A	B	C	D
28	**Income and Cash Flow Statements (Expansion)**	**2012**	**2013**	**2014**
29	Beginning-of-year Cash on Hand	NA	$15,000	$0
30	Sales (Revenue)	NA	$0	$0
31	Cost of Goods Sold	NA		
32	*Business Loan Payment*	NA		
33	Income before Taxes	NA		
34	Income Tax Expense	NA		
35	Net Income after Taxes	NA		
36	End-of-year Cash on Hand	$15,000		
37				
38	**Income and Cash Flow Statements (No Expansion)**	**2012**	**2013**	**2014**
39	Beginning-of-year Cash on Hand	NA	$15,000	$0
40	Sales (Revenue)	NA	$0	$0
41	Cost of Goods Sold	NA		
42	Income before Taxes	NA		
43	Income Tax Expense	NA		
44	Net Income after Taxes	NA		
45	End-of-year Cash on Hand	$15,000		

FIGURE C-7   Sales Revenue cells copied from the Calculations sections

The Cost of Goods Sold cells C31, D31, C41, and D41 are simply copies of the contents of cells C19, D19, C25, and D25, respectively, from the Calculations sections. Because the cells in both locations are directly below the Sales cells in the four locations, you can use the Copy command to fill those cells easily. As you can see in Figure C-8, you can drag your mouse pointer over both cells C40 and D40, right-click to see the floating toolbar, and select Copy. Move your mouse pointer to select cells C41 and D41, right-click the mouse, and select Paste. If you are uncomfortable copying and pasting with the mouse, you can type =C19, =D19, =C25, and =D25 in cells C31, D31, C41, and D41.

	A	B	C	D	E	F	G	H
28	**Income and Cash Flow Statements (Expansion)**	**2012**	**2013**	**2014**				
29	Beginning-of-year Cash on Hand	NA	$15,000	$0				
30	Sales (Revenue)	NA	$0	$0				
31	Cost of Goods Sold	NA	$0	$0				
32	*Business Loan Payment*	NA						
33	Income before Taxes	NA						
34	Income Tax Expense	NA						
35	Net Income after Taxes	NA						
36	End-of-year Cash on Hand	$15,000						
37								
38	**Income and Cash Flow Statements (No Expansion)**	**2012**	**2013**	**2014**				
39	Beginning-of-year Cash on Hand	NA	$15,000	$0				
40	Sales (Revenue)	NA	$0	$0				
41	Cost of Goods Sold	NA						
42	Income before Taxes	NA						
43	Income Tax Expense	NA						
44	Net Income after Taxes	NA						
45	End-of-year Cash on Hand	$15,000						
46								
47								
48								
49								
50								
51								
52								
53								
54								
55								
56								
57								
58								

FIGURE C-8   Cost of Goods Sold cells copied from the Calculations sections

Next you determine the Business Loan Payment for cells C32 and D32—notice that it is only present in the Income and Cash Flow Statements (Expansion) section, because if you do not expand the business, you do not need the business loan of $100,000. Excel has financial formulas to figure out loan payments. To determine a loan payment, you need to know three things: the amount being borrowed (cell C5 in the Constants section), the interest rate (cell B21 in the Calculations-Expansion section), and the number of payment periods. At the beginning of the tutorial, you learned that the bank was willing to loan money at either 5% or 4% interest compounded annually, to be paid over 10 years. Normally, banks require businesses to make monthly payments on their loans and compound the interest monthly, in which case you would enter 120 (12 months/year × 10 years) for the number of payments and divide the annual interest rate by 12 to get the period interest rate. This formula is important to remember when you enter the business world, but for now you will simplify the calculation by specifying one loan payment per year compounded annually. To put in the payment formula, click cell C32, then click the f_x symbol next to the editing window (circled in

Figure C-9). The Payment function is called PMT, so type PMT in the Insert Function window—you will immediately see a short description of the function with its arguments, as shown in Figure C-9.

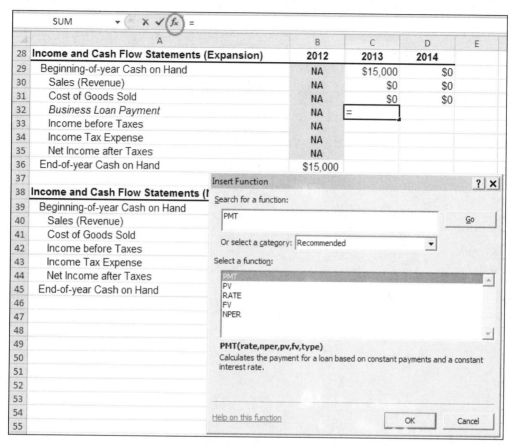

FIGURE C-9   Accessing the PMT function in Excel for cell C32

NOTE

Rate is the interest rate per period of the loan, Nper is an abbreviation for the number of loan periods, and Pv is an abbreviation for Present Value, the amount of money you are borrowing "today." The PMT function can determine a series of equal loan payments necessary to pay back the amount borrowed, plus the accumulated compound interest over the life of the loan.

When you click OK, the resulting window allows you to enter the cells or values needed in the function arguments (see Figure C-10). In the Rate text box, enter B21, which is the cell that will contain the calculated interest rate. In the Nper text box, enter 10 (for 10 years). In the Pv text box, enter C5, which is the cell that contains the loan amount.

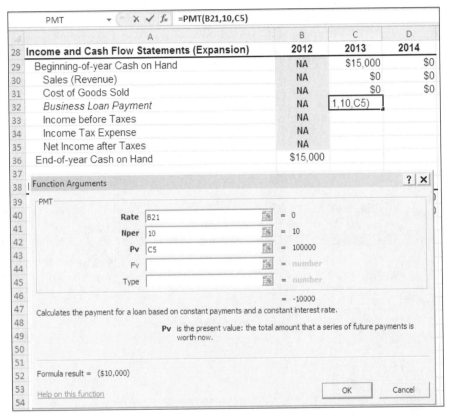

FIGURE C-10   The Function Arguments dialog box for the PMT function with the values filled in

> **NOTE**
>
> Be careful if you decide to copy the PMT formula from cell C32 into cell D32, because the Copy command will change the cells in the formula arguments to the next adjacent cells. To make the Copy command work correctly, you have two options. First, you can change the Rate and Pv cells in the cell C32 formula from *relative reference* (B21, C5) to *absolute reference* ($B$21, $C$5). Your other option is to re-insert the PMT function into cell D32 and type the same arguments as before in the boxes. Absolute referencing of a cell (using $ signs in front of the Column and Row designators) "anchors" the cell so that when the Copy command is used, the destination cell will refer back to the same cells that the source cell used. If necessary, consult the Excel online Help for an explanation of relative and absolute cell references.

When you click OK ($10,000) should appear in cell C32. Payments in Excel always appear as negative numbers, which is why the number has parentheses around it. (Depending on your cell formatting, the number may also appear in red.) Next, you need to have the same payment amount in cell D32 (for 2014). Because the PMT function creates equal payments over the life of the loan, you can simply type =C32 into cell D32.

The next line in the Income and Cash Flow Statements is Income before Taxes, which is an easy calculation. It is the Sales minus the Cost of Goods Sold, minus the Business Loan Payment. However, because the PMT function shows the loan payment as a negative number, you will instead add the Business Loan Payment. In cell C33, enter =C30-C31+C32. Again, a negative $10,000 should be displayed, as the cells other than the loan payment currently have zero in them. Copy cell C33 to cell D33. In cell C42 of the next section below (No Expansion), enter =C40-C41. (There is no loan payment in this section to put in the calculation.) Next, copy cell C42 to cell D42. At this point, your Income and Cash Flow Statements should look like Figure C-11.

	D42	f_x =D40-D41			
	A		B	C	D
28	**Income and Cash Flow Statements (Expansion)**		**2012**	**2013**	**2014**
29	Beginning-of-year Cash on Hand		NA	$15,000	$0
30	Sales (Revenue)		NA	$0	$0
31	Cost of Goods Sold		NA	$0	$0
32	*Business Loan Payment*		NA	($10,000)	($10,000)
33	Income before Taxes		NA	-$10,000	-$10,000
34	Income Tax Expense		NA		
35	Net Income after Taxes		NA		
36	End-of-year Cash on Hand		$15,000		
37					
38	**Income and Cash Flow Statements (No Expansion)**		**2012**	**2013**	**2014**
39	Beginning-of-year Cash on Hand		NA	$15,000	$0
40	Sales (Revenue)		NA	$0	$0
41	Cost of Goods Sold		NA	$0	$0
42	Income before Taxes		NA	$0	$0
43	Income Tax Expense		NA		
44	Net Income after Taxes		NA		
45	End-of-year Cash on Hand		$15,000		

FIGURE C-11   The Income and Cash Flow Statements completed up to Income before Taxes

Income Tax Expense is the most complex formula for these sections. Because you do not pay income tax when you have no income or a loss, you must use a formula that allows you to enter 0 if there is no income or a loss, or to calculate the tax rate on a positive income. You can use the IF function in Excel to enter one of two results in a cell, depending on whether a defined logical statement is true or false. To create an IF function, select cell C34, then click the f_x symbol next to the cell editing window (circled in Figure C-12). When the Insert Function dialog box appears, type IF in the "Search for a function" text box, and click the Go button if necessary. The IF function should appear. When you click OK, the Function Arguments dialog box appears (see Figure C-13).

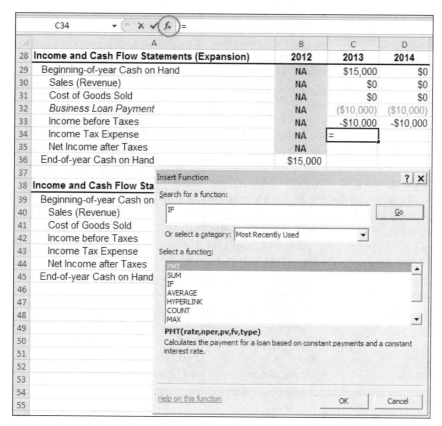

FIGURE C-12   The IF function

Type the following in the Function Arguments dialog box:

- Next to Logical_test, type C33<=0.
- Next to Value_if_true, type 0.
- Next to Value_if_false, type C33*C4 (the Income before Taxes multiplied by the Tax Rate for 2013).

As you fill in the arguments, Excel writes the formula for you in the formula editing window (circled in Figure C-13).

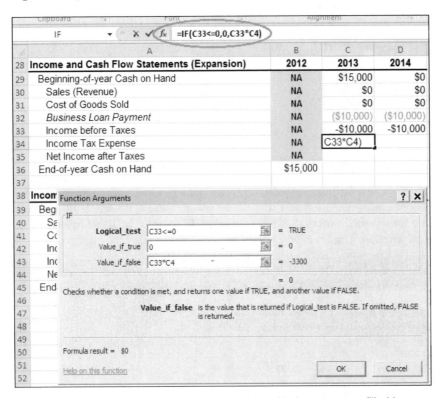

FIGURE C-13   The Function Arguments dialog box with the arguments filled in

Once you have entered the arguments, click OK; Excel enters the formula into the cell. Because you had negative income, the cell should display a zero for now. Because the same formula will be used in 2014 (but with the 2014 tax rate), you can simply copy and paste the formula from cell C34 to cell D34. You also have to calculate the income tax for the Income and Cash Flow Statements (No Expansion). In cell D43, use the same IF function, but in the Logical_test, Value_if_true, and Value_if_false arguments, you must type C42<=0, 0, and C42*C4, respectively. Again, the cell will display $0 for an answer. Copy cell C43 to cell D43 to complete the Income Tax Expense line for No Expansion.

Net Income after Taxes is simply the Income before Taxes minus the Income Tax Expense. Enter the formula into cell C35, then copy cell C35 over to cells D35, C44, and D44. If you did this correctly, cells C35 and D35 will display a negative $10,000, and cells C44 and D44 will display $0.

End-of-year Cash on Hand, the last line in both Income and Cash Flow Statements sections, is not difficult either. Conceptually, the cash you have at the end of the year is equal to your Beginning-of-year Cash on Hand plus your Net Income after Taxes. Enter the formula into cell C36, then copy cell C36 over to cell D36. Note that because the Income and Cash Flow Statements (No Expansion) do not have a line item for Business Loan Payment, you cannot copy the same command down to it. You have to enter the formula manually for cell C45, which is =C39+C44. However, you can copy cell C45 to cell D45 to finish the Income and Cash Flow Statements sections. The completed sections should look like Figure C-14.

	A	B	C	D
	D45	▾	f_x =D39+D44	
28	**Income and Cash Flow Statements (Expansion)**	**2012**	**2013**	**2014**
29	Beginning-of-year Cash on Hand	NA	$15,000	$5,000
30	Sales (Revenue)	NA	$0	$0
31	Cost of Goods Sold	NA	$0	$0
32	*Business Loan Payment*	NA	($10,000)	($10,000)
33	Income before Taxes	NA	-$10,000	-$10,000
34	Income Tax Expense	NA	$0	$0
35	Net Income after Taxes	NA	-$10,000	-$10,000
36	End-of-year Cash on Hand	$15,000	$5,000	-$5,000
37				
38	**Income and Cash Flow Statements (No Expansion)**	**2012**	**2013**	**2014**
39	Beginning-of-year Cash on Hand	NA	$15,000	$15,000
40	Sales (Revenue)	NA	$0	$0
41	Cost of Goods Sold	NA	$0	$0
42	Income before Taxes	NA	$0	$0
43	Income Tax Expense	NA	$0	$0
44	Net Income after Taxes	NA	$0	$0
45	End-of-year Cash on Hand	$15,000	$15,000	$15,000

FIGURE C-14   The completed Income and Cash Flow Statements sections

Filling in the "Hard" Formulas

To finish the spreadsheet, you will enter values in the Inputs section and write the formulas in both Calculations sections.

AT THE KEYBOARD

In cell C8, enter an R for Recession, and in cell C9, enter H for High Inflation. You could enter any values here, but these two values will work with the IF functions you will write later. Recall that you did not use separate inputs for 2013 and 2014. You are assuming that the economic outlook or inflation rate that exists for 2013 will extend into 2014. However, because you are using the same inputs from these two locations, you must remember to use *absolute* cell references to both cells C8 and C9 in the various IF statements if you want to use a Copy command for adjacent cells. Your Inputs section should look like the one in Figure C-15.

	A	B	C	D
7	**Inputs**	**2012**	**2013**	**2014**
8	Economic Outlook (R=Recession, B=Boom)	NA	R	NA
9	Inflation Outlook (H=High, L=Low)	NA	H	NA

FIGURE C-15   The Inputs section with values entered in cells C8 and C9

Remember that you referred to cell addresses in both Calculations sections in your formulas in the Income and Cash Flow Statements sections. Now you will enter formulas for these calculations. If necessary, format the four Total Sales Dollars cells and the four Cost of Goods Sold cells in the Calculations sections as Currency with no decimal places.

As described at the beginning of the tutorial, the forecast for Total Sales Dollars is a function of both the Economic Outlook and whether you expand the business. The following table lists the predicted sales growth percentages:

Sales Growth Forecast—Collegetown Thrift Shop

	Business Expansion	**No Business Expansion**
Recession-R	30%	20%
Boom-B	15%	5%

You will use IF formulas to forecast Total Sales Dollars. Click cell C18, then bring up the IF function and type the following in the text boxes:

Logical_test: $C$8="R" (Note that you must use absolute cell referencing for cell B8 and quotation marks for Excel to recognize a text string.)

Value_if_true: B18*1.3 (the 2012 sales multiplied by 1.3 for 30% sales growth)

Value_if_false: B18*1.15 (the 2012 sales multiplied by 1.15 for 15% sales growth)

Compare your entries to Figure C-16.

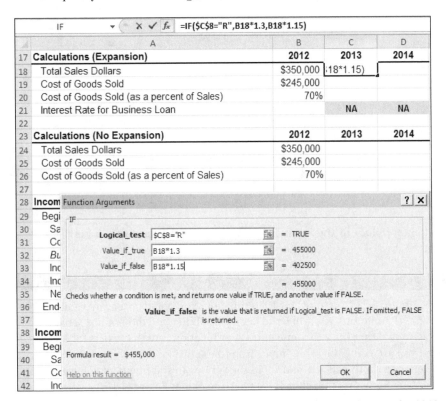

FIGURE C-16   Using the IF function to enter the Total Sales Dollars forecast for 2013

When you click OK, cell C18 should display $455,000, because 30% of $350,000 is $105,000, and $350,000 plus $105,000 equals $455,000. So, it appears that the formula returned a "true" value with an R inserted in cell C8. Because you "anchored" cell C8 by entering $C$8, copy this formula over to cell D18 for the year 2014.

Once you complete the Total Sales Dollars cells for the Expansion scenario, go down to the Calculations (No Expansion) section and use IF statements to enter formulas for the Total Sales Dollars. Use a 20% sales growth factor for Recession and 5% for Boom. You can copy the formula from cell C18 into cell C24, but you then will have to use the editing window to change the values in the true and false arguments from 1.3 and 1.15 to 1.2 and 1.05, respectively, to reflect the fact that you did not expand the business. See Figure C-17.

| C24 | fx | =IF($C$8="R",B24*1.2,B24*1.05) |

	A	B	C	D
17	**Calculations (Expansion)**	**2012**	**2013**	**2014**
18	Total Sales Dollars	$350,000	$455,000	$591,500
19	Cost of Goods Sold	$245,000		
20	Cost of Goods Sold (as a percent of Sales)	70%		
21	Interest Rate for Business Loan		NA	NA
22				
23	**Calculations (No Expansion)**	**2012**	**2013**	**2014**
24	Total Sales Dollars	$350,000	$420,000	$504,000
25	Cost of Goods Sold	$245,000		
26	Cost of Goods Sold (as a percent of Sales)	70%		
27				
28	The formula was copied from cell C18, then the values			
29	in the arguments were changed to 1.2 and 1.05			

FIGURE C-17   Copying cell C18 into cell C24 and then editing the IF function arguments to change the sales growth percentages

As before, you can now copy cell C24 to cell D24. You have completed the Total Sales Dollars calculations.

The Cost of Goods Sold (cells C19, D19, C25, and D25) is the Total Sales Dollars multiplied by the Cost of Goods Sold as a percent of Sales. In cell C19, type =C18*C20 and press Enter. Copy cell C19 and paste the contents into cells D19, C25, and D25. Your answers will be $0 until you enter the formulas for the Cost of Goods Sold as a percent of Sales.

The Cost of Goods Sold as a percent of Sales (cells C20, D20, C26, and D26) was 70% in 2012. In variety merchandising for resold items, it is easier to use an aggregate measure such as Cost of Goods Sold as a percent of Sales rather than trying to capture an individual Cost of Goods Sold for each item. From the 2012 data, you determined that for every dollar of sales you collected in 2012, you spent 70 cents purchasing the item and preparing it for resale. You will use that percentage as a basis for forecasting Cost of Goods Sold as a percent of Sales, applying an appropriate inflation factor for the cost of acquiring the stock for sale. The following table lists the predicted inflation percentages for Cost of Goods Sold.

Cost of Goods Sold Forecast—Collegetown Thrift Shop

	Business Expansion	**No Business Expansion**
High Inflation	6%	6%
Low Inflation	2%	2%

As with Total Sales Dollars previously, you will again use the IF function to calculate the Cost of Goods Sold as a percent of Sales. Now that you are familiar with the IF function, you can probably enter the function without using the dialog boxes. In cell C20, type the following:

=IF($C$9="H",B20*1.06,B20*1.02)

This expression means that if the text string in cell C9 is the letter H, you multiply the value in cell B20 by 1.06 (6% inflation). If the value in cell C9 is not an H, multiply the value in cell B20 by 1.02 (2% inflation). The value in cell B20 was the baseline Cost of Goods Sold as a percent of Sales in 2012, which was 70%. You can now copy cell C20 and paste the contents into cell D20.

Because the inflation percentages were exactly the same for both the Expansion and No Expansion calculations, you can also copy cell C20 and paste the contents into cells C26 and D26. Your Calculations sections should now look like Figure C-18.

D26		fx	=IF($C$9="H",C26*1.06,C26*1.02)	

	A	B	C	D
		2012	2013	2014
17	**Calculations (Expansion)**			
18	Total Sales Dollars	$350,000	$455,000	$591,500
19	Cost of Goods Sold	$245,000	$337,610	$465,227
20	Cost of Goods Sold (as a percent of Sales)	70%	74%	79%
21	Interest Rate for Business Loan		NA	NA
22				
23	**Calculations (No Expansion)**	2012	2013	2014
24	Total Sales Dollars	$350,000	$420,000	$504,000
25	Cost of Goods Sold	$245,000	$311,640	$396,406
26	Cost of Goods Sold (as a percent of Sales)	70%	74%	79%

FIGURE C-18   Calculations sections nearly complete

The last item in the Calculations section is the Interest Rate for Business Loan (cell B21). Remember the bank's statement that if the economy recovers, it could lower the interest rate from 5% to 4%. So, you will need one more IF function to insert into cell B21 based on the economic outlook. If the economic outlook is for a Recession (R), then the interest rate will be 5% annually; if the outlook is for a Boom (B), then the interest rate will be 4% annually. Now that you are familiar with the IF function, you can simply type the expression into the cell yourself. Click cell B21, type =IF(C8="R",5%,4%), and press Enter.

You will immediately notice that 5% appears in the cell because you have R in the input cell for Economic Outlook. You may also notice that you now have a negative $12,950 in the Business Loan Payment cells (C32 and D32). See Figure C-19 to compare your results.

	A	B	C	D
		2012	2013	2014
7	**Inputs**			
8	Economic Outlook (R=Recession, B=Boom)	NA	R	NA
9	Inflation Outlook (H=High, L=Low)	NA	H	NA
10				
11	**Summary of Key Results**	2012	2013	2014
12	Net Income after Taxes (Expansion)	NA	$69,974	$73,660
13	End-of-year Cash on Hand (Expansion)	NA	$84,974	$158,634
14	Net Income after Taxes (No Expansion)	NA	$72,601	$69,936
15	End-of-year Cash on Hand (No Expansion)	NA	$87,601	$157,537
16				
17	**Calculations (Expansion)**	2012	2013	2014
18	Total Sales Dollars	$350,000	$455,000	$591,500
19	Cost of Goods Sold	$245,000	$337,610	$465,227
20	Cost of Goods Sold (as a percent of Sales)	70%	74%	79%
21	Interest Rate for Business Loan	5%	NA	NA
22				
23	**Calculations (No Expansion)**	2012	2013	2014
24	Total Sales Dollars	$350,000	$420,000	$504,000
25	Cost of Goods Sold	$245,000	$311,640	$396,406
26	Cost of Goods Sold (as a percent of Sales)	70%	74%	79%
27				
28	**Income and Cash Flow Statements (Expansion)**	2012	2013	2014
29	Beginning-of-year Cash on Hand	NA	$15,000	$84,974
30	Sales (Revenue)	NA	$455,000	$591,500
31	Cost of Goods Sold	NA	$337,610	$465,227
32	Business Loan Payment	NA	($12,950)	($12,950)
33	Income before Taxes	NA	$104,440	$113,323
34	Income Tax Expense	NA	$34,465	$39,663
35	Net Income after Taxes	NA	$69,974	$73,660
36	End-of-year Cash on Hand	$15,000	$84,974	$158,634
37				
38	**Income and Cash Flow Statements (No Expansion)**	2012	2013	2014
39	Beginning-of-year Cash on Hand	NA	$15,000	$87,601
40	Sales (Revenue)	NA	$420,000	$504,000
41	Cost of Goods Sold	NA	$311,640	$396,406
42	Income before Taxes	NA	$108,360	$107,594
43	Income Tax Expense	NA	$35,759	$37,658
44	Net Income after Taxes	NA	$72,601	$69,936
45	End-of-year Cash on Hand	$15,000	$87,601	$157,537

FIGURE C-19   The finished spreadsheet

You can change the economic inputs in four different combinations: R-H, R-L, B-H, B-L. This allows you to see the impact on your net income and cash on hand both for expanding and not expanding. However, you have another more powerful way to do this. In the next section, you will learn how to tabulate the financial results of the four possible combinations using an Excel tool called Scenario Manager.

SCENARIO MANAGER

You are now ready to evaluate the four possible outcomes for your DSS model. Because this is a simple, four-outcome model, you could have created four different spreadsheets, one for each set of outcomes, and then transferred the financial information from each spreadsheet to a Summary Report.

In essence, Scenario Manager performs the same task. It runs the model for all the requested outcomes and presents a tabular summary of the results. This summary is especially useful for reports and presentations needed by upper managers, financial investors, or in this case, the bank.

To review, the four possible combinations of input values are: R-H (Recession and High Inflation), R-L (Recession and Low Inflation), B-H (Boom and High Inflation), and B-L (Boom and Low Inflation). You could consider each combination of inputs a separate scenario. For each of these scenarios, you are interested in four outputs: Net Income after Taxes for Expansion and No Expansion, and End-of-year Cash on Hand for Expansion and No Expansion.

Scenario Manager runs each set of combinations and then records the specified outputs as a summary into a separate worksheet. You can use these summary values as a table of numbers and print it, or you can copy them into a Microsoft Word document or a PowerPoint presentation. You can also use the data table to build a chart or graph, which you can put into a report or presentation.

When you define a scenario in Scenario Manager, you name it and identify the input cells and input values. Then you identify the output cells so Scenario Manager can capture the outputs in a summary sheet.

AT THE KEYBOARD

To start, click the Data tab on the Ribbon. In the Data Tools group, click the What-If Analysis button, then click Scenario Manager from the menu that appears (see Figure C-20).

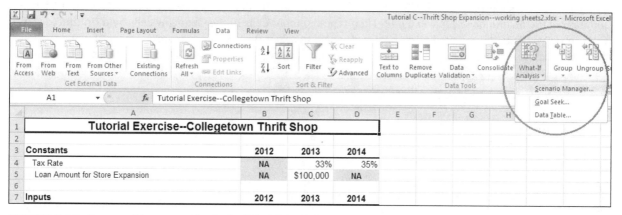

FIGURE C-20   Scenario Manager option in the What-If Analysis menu

Scenario Manager appears (see Figure C-21), but no scenarios are defined. Use the dialog box to add, delete, or edit scenarios.

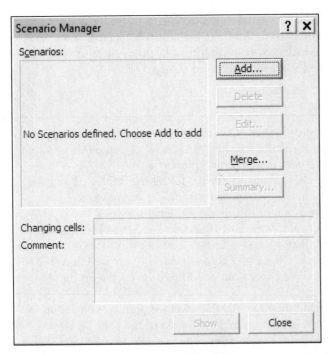

FIGURE C-21   Initial Scenario Manager dialog box

> **NOTE**
>
> When working with the Scenario Manager dialog box and any following dialog boxes, do not use the Enter key to navigate. Use mouse clicks to move from one step to the next.

To define a scenario, click the Add button. The Edit Scenario dialog box appears. Enter Recession-High Inflation in the field under Scenario name. Then type the input cells in the Changing cells field (in this case, C8:C9). Better yet, you can use the button next to the field to select the cells in your spreadsheet. If you do, Scenario Manager changes the cell references to absolute cell references, which is acceptable (see Figure C-22).

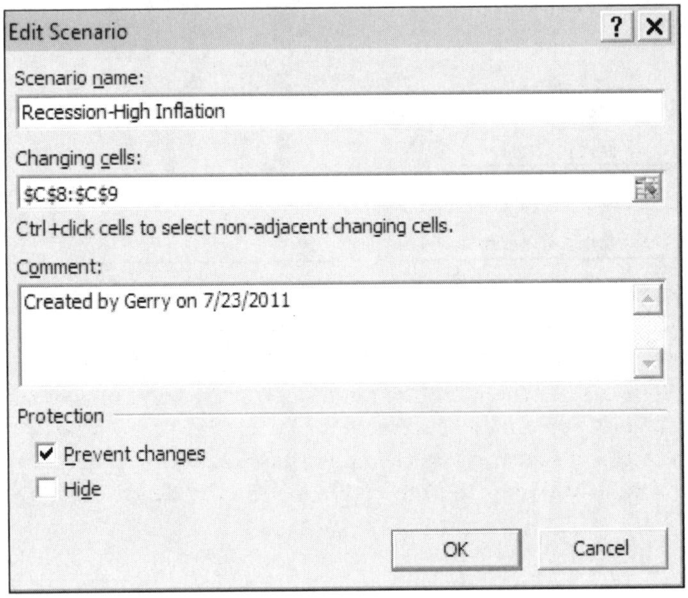

FIGURE C-22   Defining a scenario name and input cells

Click OK to open the Scenario Values dialog box. Enter the input values for the scenario. In the case of Recession and High Inflation, the values will be R and H for cells C8 and C9, respectively (see Figure C-23). Note that if you already have entered values in the spreadsheet, the dialog box will display the current values. Make sure to enter the correct values.

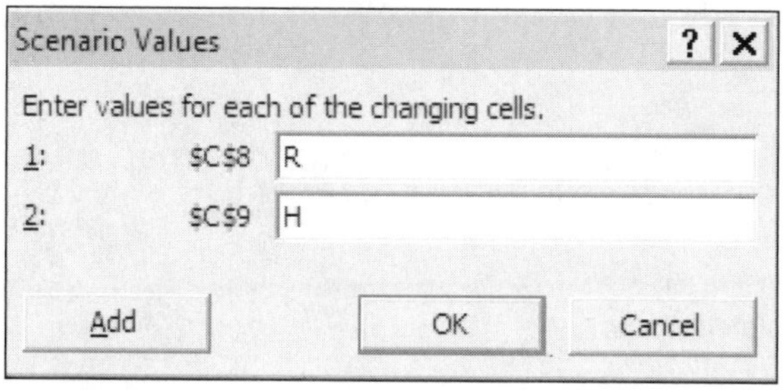

FIGURE C-23 Entering values for the input cells

Click OK to return to the Scenario Manager dialog box. Enter the other three scenarios: Recession-Low Inflation, Boom-High Inflation, and Boom-Low Inflation (R-L, B-H, and B-L), and their related input values. When you finish, you should see the names and changing cells for the four scenarios (see Figure C-24).

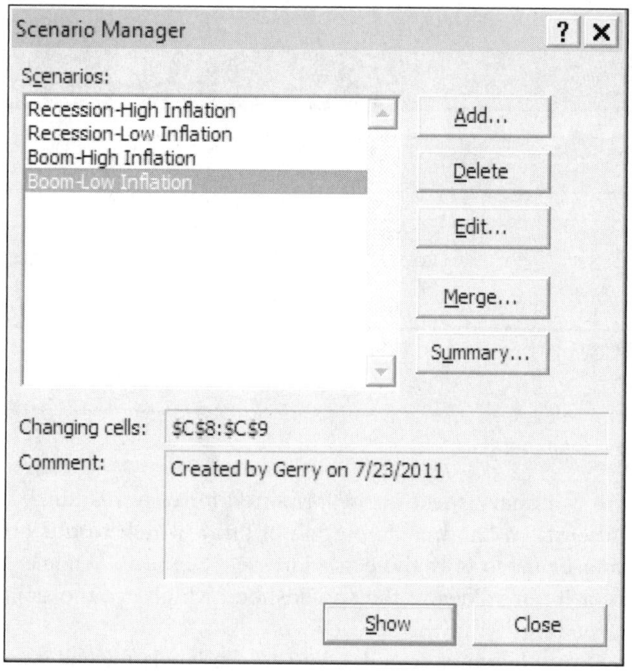

FIGURE C-24 Scenario Manager dialog box with all four scenarios entered

You can now create a summary sheet that displays the results of running the four scenarios. Click the Summary button to open the Scenario Summary dialog box, as shown in Figure C-25. You must now enter the output cell addresses in Excel—they will be the same for all four scenarios. Recall that you created a section in your spreadsheet called Summary of Key Results. You are primarily interested in the results at the end of 2014, so you will choose the four cells that represent the Net Income after Taxes and End-of-year Cash on Hand, and then use them for both the expansion scenario and the non-expansion scenario. These cells are D12 to D15 in your spreadsheet. Either type D12:D15 or use the button next to the Result cells field and select those cells in the spreadsheet.

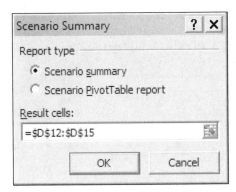

FIGURE C-25   Scenario Summary dialog box with Result cells entered

Another good reason for having a Summary of Key Results section is that it provides a contiguous range of cells to define for summary output. However, if you want to add output from other cells in the spreadsheet, simply separate each cell or range of cells in the dialog box with a comma. Next, click OK. Excel runs each set of inputs in the background, collects the results from the result cells, and then creates a new sheet called Scenario Summary (the name on the sheet's lower tab), as shown in Figure C-26.

			Current Values:	Recession-High Inflation	Recession-Low Inflation	Boom-High Inflation	Boom-Low Inflation
Scenario Summary							
Changing Cells:							
$C$8	R		R	R	B	B	
$C$9	H		H	L	H	L	
Result Cells:							
$D$12			$73,660	$73,660	$96,052	$56,216	$73,738
$D$13			$158,634	$158,634	$189,562	$132,531	$157,605
$D$14			$69,936	$69,936	$89,015	$53,545	$68,152
$D$15			$157,537	$157,537	$184,496	$132,071	$153,573

Notes: Current Values column represents values of changing cells at time Scenario Summary Report was created. Changing cells for each scenario are highlighted in gray.

FIGURE C-26   Scenario Summary sheet created by Scenario Manager

As you can see, the output created by the Scenario Summary sheet is not formatted for easy reading. You do not know which results are the net income and cash on hand, and you do not know which results are for Expansion vs. No Expansion, because Scenario Manager listed only the cell addresses. Scenario Manager also listed a separate column (column D) for the current input values in the spreadsheet, which are the same as the values in column E. It also left a blank column (column A) in the spreadsheet.

Fortunately, it is fairly easy to format the output. Delete columns D and A, put in the labels for cell addresses in the new column A, and then retitle the Scenario Summary as Collegetown Thrift Shop Financial Forecast, End of Year 2014 (because you are looking only at Year 2014 results). You can also make the results columns narrower by breaking the column headings into two lines; place your cursor in the editing window where you want to break the words, and then press Alt+Enter. Add a heading for column B (Cell Address). Finally, merge and center the title, and center the column headings and the input cell values (R, B, H, and L). Leave your financial data right-justified to keep the numbers lined up correctly. Finally, put some border boxes around each column of results. Figure C-27 shows a formatted Scenario Summary worksheet.

		Cell Address	Recession-High Inflation	Recession-Low Inflation	Boom-High Inflation	Boom-Low Inflation
2	Scenario Summary--Collegetown Thrift Shop Financial Forecast, End of Year 2014					
5	**Changing Cells:**					
6	Economic Outlook: R-Recession, B-Boom	$C$8	R	R	B	B
7	Inflation: H-High, L-Low	$C$9	H	L	H	L
8	**Result Cells:**					
9	Net Income after Taxes (Expansion)	$D$12	$73,660	$96,052	$56,216	$73,738
10	End-of-year Cash on Hand (Expansion)	$D$13	$158,634	$189,562	$132,531	$157,605
11	Net Income after Taxes (No Expansion)	$D$14	$69,936	$89,015	$53,545	$68,152
12	End-of-year Cash on Hand (No Expansion)	$D$15	$157,537	$184,496	$132,071	$153,573

13 Notes: Current Values column represents values of changing cells at
14 time Scenario Summary Report was created. Changing cells for each
15 scenario are highlighted in gray.

FIGURE C-27   Scenario Summary worksheet after formatting and adding labels

Interpreting the Results

Now that you have good data, what do you do with it? Remember, you wanted to see if taking a $100,000 business loan to expand the thrift shop was a good financial decision. This is a relatively simple business case, and the shop's success so far ($350,000 of sales in 2012) would seem to make expansion a good risk. But how good a risk is the expansion?

After building the spreadsheet and doing the analysis, you can make comparisons and interpret the results. Regardless of the economic outlook or inflation, all four scenarios indicate that expanding the business should provide greater Net Income After Taxes and End-of-year Cash on Hand (after two years) than not expanding. So, the DSS model not only provides a quantitative basis for expanding, it provides an analysis that you can present to prospective lenders.

What decision would you make about expansion if you looked only at the 2013 forecast? You could go back to the original spreadsheet and look at the figures for 2013, or you can go to Scenario Manager and create a new summary, specifying the 2013 cells C12 through C15. See Figure C-28.

	A	B 2012	C 2013	D 2014
7	**Inputs**	2012	2013	2014
8	Economic Outlook (R=Recession, B=Boom)	NA	R	NA
9	Inflation Outlook (H=High, L=Low)	NA	H	NA
10				
11	**Summary of Key Results**	2012	2013	2014
12	Net Income after Taxes (Expansion)	NA	$69,974	$73,660
13	End-of-year Cash on Hand (Expansion)	NA	$84,974	$158,634
14	Net Income after Taxes (No Expansion)	NA	$72,601	$69,936
15	End-of-year Cash on Hand (No Expansion)	NA	$87,601	$157,537
16				
17	**Calculations (Expansion)**	2012	2013	2014
18	Total Sales Dollars	$350,000	$455,000	$591,500
19	Cost of Goods Sold	$245,000	$337,610	$465,227
20	Cost of Goods Sold (as a percent of Sales)	70%	74%	79%

Scenario Summary dialog: Report type — Scenario summary / Scenario PivotTable report; Result cells: =$C$12:$C$15; OK / Cancel

FIGURE C-28   Creating a new Scenario Summary for 2013 instead of 2014

When you click OK, Excel creates a second Scenario Summary (appropriately named Scenario Summary 2), but this time the output values come from 2013, not 2014. After editing and formatting, the 2013 Scenario Summary should look like Figure C-29.

	A	B	C	D	E	F
1						
2	Scenario Summary 2--Collegetown Thrift Shop Financial Forecast--End of Year 2013					
3		Cell Address	Recession-High Inflation	Recession-Low Inflation	Boom-High Inflation	Boom-Low Inflation
5	Changing Cells:					
6	Economic Outlook: R-Recession, B-Boom	$C$8	R	R	B	B
7	Inflation: H-High, L-Low	$C$9	H	L	H	L
8	Result Cells:					
9	Net Income after Taxes (Expansion)	$C$12	$69,974	$78,510	$61,316	$68,867
10	End-of-year Cash on Hand (Expansion)	$C$13	$84,974	$93,510	$76,316	$83,867
11	Net Income after Taxes (No Expansion)	$C$14	$72,601	$80,480	$63,526	$70,420
12	End-of-year Cash on Hand (No Expansion)	$C$15	$87,601	$95,480	$78,526	$85,420
13	Notes: Current Values column represents values of changing cells at					
14	time Scenario Summary Report was created. Changing cells for each					
15	scenario are highlighted in gray.					

FIGURE C-29   Scenario Summary for End of Year 2013

As you can see, *not* expanding the business yields slightly better financial results at the end of 2013. As the original Scenario Summary points out, it will take two years for the business expansion to start making more money when compared with not expanding. You can also revise the original spreadsheet to copy the columns out to 2014, 2015, and beyond to forecast future income and cash flows. However, note that the accuracy of a forecast gets worse as you extend it in time.

Managers must also maintain a healthy skepticism about the validity of their assumptions when formulating a DSS model. Most assumptions about economic outlooks, inflation, and interest rates are really educated guesses. For example, who could have predicted the economic meltdown in 2007? Business DSS models for investments, new product launches, business expansion, or major capital projects commonly look at three possible outcomes: best case, most likely, and worst case. The most likely outcome is based on previous years' data already collected by the firm. The best-case and worst-case outcomes are formulated based on some percentage of performance that falls above or below the most likely scenario. At least these are data-driven forecasts, or what people in the business world call "guessing—with data."

So, how do you reduce risk when making financial decisions based on DSS model results? It helps to formulate the model based on valid data and to use conservative estimates for success. More importantly, collecting pertinent data and tracking the business results *after* deciding to invest or expand can help reduce the risk of failure for the enterprise.

Summary Sheets

When you start working on the Scenario Manager spreadsheet cases later in this book, you will need to know how to manipulate summary sheets and their data. Some of these operations are explained in the following sections.

Rerunning Scenario Manager

The Scenario Summary sheet does not update itself when you change formulas or inputs in the spreadsheet. To get an updated Scenario Summary, you must rerun Scenario Manager, as you did when changing the outputs from 2014 to 2013. Click the Summary button in the Scenario Manager dialog box, and then click OK. Another summary sheet is created; Excel numbers them sequentially (Scenario Summary, Scenario Summary2, etc.), so you do not have to worry about Excel overwriting any of your older summaries. That is why you should rename each summary with a description of the changes.

Deleting Unwanted Scenario Manager Summary Sheets

When working with Scenario Manager, you might produce summary sheets you do not want. To delete an unwanted sheet, move your mouse pointer to the group of sheet tabs at the bottom of the screen and *right*-click the tab of the sheet you want to delete. Click Delete from the menu that appears (see Figure C-30).

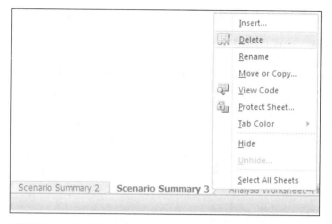

FIGURE C-30   Deleting unwanted worksheets

Charting Summary Sheet Data

You can easily chart Summary Sheet results using the Charts group in the Insert tab, as discussed in Tutorial F. Figure C-31 shows a clustered column chart prepared from the data in the Scenario Summary for 2014. Charts are useful because they provide a visual comparison of results. As the chart shows, the best economic climate for the thrift shop is a Recession with Low Inflation.

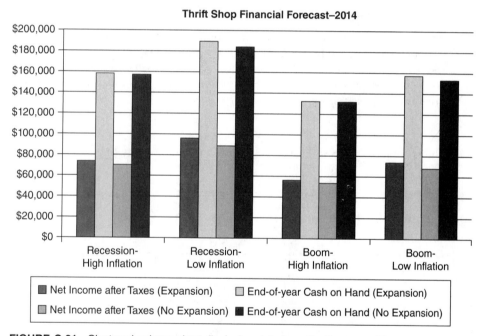

FIGURE C-31   Clustered column chart displaying data from the Summary Sheet

Copying Summary Sheet Data to the Clipboard

As you can with almost everything else in Microsoft Office, you can copy summary sheet data to other Office applications (a Word document or PowerPoint slide, for example) by using the Clipboard. Follow these steps:

1. Select the data range you want to copy.
2. Right-click the mouse and select Copy from the resulting menu.
3. Open the Word document or PowerPoint presentation into which you want to copy.
4. Click your cursor where you want the upper-left corner of the copied data to be displayed.
5. Right-click the mouse and select Paste from the resulting menu. The data should now appear on your document.

PRACTICE EXERCISE—TED AND ALICE'S HOUSE PURCHASE DECISION

Ted and Alice are a young couple who have been living in an apartment for the first two years of their marriage. They would like to buy their first house, but do not know whether they can afford it. Ted works as a carpenter's apprentice, and Alice is a customer service specialist at a local bank. In 2012, Ted's "take home" wages were $24,000 after taxes and deductions, and Alice's take-home salary was $30,000. Ted gets a 2% raise every year, and Alice gets a 3% raise. Their apartment rent is $1,200 per month ($14,400 per year), but the lease is up for renewal and the landlord said he needs to increase the rent for the next lease.

Ted and Alice have been looking at houses and have found one they can buy, but they will need to borrow $200,000 for a mortgage. Their parents are helping them with the down payment and closing costs. After talking to several lenders, Ted and Alice have learned that the state legislature is voting on a first-time home buyers' mortgage bond. If the bill passes, they will be able to get a 30-year fixed mortgage at 3% interest. Otherwise, they will have to pay 6% interest on the mortgage.

Because of the depressed housing market, Ted and Alice are not figuring equity value into their calculations. In addition, although the mortgage interest and real estate taxes will be deductible on their income taxes, these deductions will not be higher than the standard allowable tax deduction, so they are not figuring on any savings there either. Ted and Alice's other living expenses (such as car payments, food, and medical bills), the utilities expenses for either renting or buying, and estimated house maintenance expenses are listed in the Constants section (see Figure C-32).

Ted and Alice's primary concern is their cash on hand at the end of years 2013 and 2014. They are thinking of starting a family, but they know it will be difficult without adequate savings.

Getting Started on the Practice Exercise

If you closed Excel after the first tutorial exercise, start Excel again—it will automatically open a new workbook for you. If your Excel workbook from the first tutorial is still open, you may find it useful to start a new worksheet in the same workbook. Then you can refer back to the first tutorial when you need to structure or format the spreadsheet; the formatting of both exercises in this tutorial is similar. Set up your new worksheet as explained in the following sections.

Constants Section

Your spreadsheet should have the constants shown in Figure C-32. An explanation of the line items follows the figure.

	A	B	C	D
1	Tutorial Exercise Skeleton--Ted and Alice's House Decision			
2				
3	Constants	2012	2013	2014
4	Non-Housing Living Expenses (Cars, Food, Medical, etc)	NA	$36,000	$39,000
5	Mortgage Amount for Home Purchase	NA	$200,000	NA
6	Real Estate Taxes and Insurance on Home	NA	$3,000	$3,150
7	Utilities Expense (Heat & Electric)--Apartment	NA	$2,000	$2,200
8	Utilities Expense(Heat, Electric, Water, Trash)--House	NA	$2,500	$2,600
9	House Repair and Maintenance Expenses	NA	$1,200	$1,400

FIGURE C-32   Constants section

- Non-Housing Living Expenses—This value represents Ted and Alice's estimate of all their other living expenses for 2013 and 2014.
- Mortgage Amount for Home Purchase
- Real Estate Taxes and Insurance on Home—A lender has given Ted and Alice estimates for these values; they are usually paid monthly with the house mortgage payment. The money is placed in an escrow account and then paid by the mortgage company to the state or county and insurance company.
- Utilities Expense—Apartment—This value is Ted and Alice's estimate for 2013 and 2014 based on their 2012 bills.

- Utilities Expense—House—Currently the apartment rent includes fees for water, sewer, and trash disposal. If they get a house, Ted and Alice expect the utilities to be higher.
- House Repair and Maintenance Expenses—In an apartment, the landlord is responsible for repair and maintenance. Ted and Alice will have to budget for repair and maintenance on the house.

Inputs Section

Your spreadsheet should have the inputs shown in Figure C-33. An explanation of line items follows the figure.

	A	B	C	D
11	Inputs	2012	2013	2014
12	Rental Occupancy (H=High, L=Low)	NA		NA
13	First Time Buyer Bond Loans Available (Y=Yes, N=No)	NA		NA

FIGURE C-33 Inputs section

- Rental Occupancy (H=High, L=Low)—When the housing market is depressed (in other words, people are not buying homes), rental housing occupancy percentages are high, which allows landlords to charge higher rents when leases are renewed. Ted and Alice think their rent will increase in 2013. The amount of the increase depends on the Rental Occupancy. If the occupancy is high, Ted and Alice expect to see a 10% increase in rent in both 2013 and 2014. If occupancy is low, they only expect a 3% increase.
- First Time Buyer Bond Loans Available (Y=Yes, N=No)—As described earlier, when housing markets are depressed, local governments will frequently pass a bond bill to provide low-interest mortgage money to first-time home buyers. If the bond loans are available, Ted and Alice can obtain a 30-year fixed mortgage at only 3%, which is half the interest rate they would otherwise pay for a conventional mortgage.

Summary of Key Results Section

Figure C-34 shows what key results Ted and Alice are looking for. They want to know their End-of-year Cash on Hand for both 2013 and 2014 if they decide to stay in the apartment and if they decide to purchase the house.

	A	B	C	D
15	Summary of Key Results	2012	2013	2014
16	End-of-year Cash on Hand (Rent)	NA		
17	End-of-year Cash on Hand (Buy)	NA		

FIGURE C-34 Summary of Key Results section

These results are copied from the End-of-year Cash on Hand sections of the Income and Cash Flow Statements sections (for both renting and buying).

Calculations Section

Your spreadsheet will need formulas to calculate the apartment rent, house payments, and interest rate for the mortgage (see Figure C-35). You will use the rent and house payments later in the Income and Cash Flow Statements for both renting and buying.

	A	B	C	D
19	Calculations	2012	2013	2014
20	Apartment Rent	$14,400		
21	House Payments	NA		
22	Interest Rate for House Mortgage		NA	NA

FIGURE C-35 Calculations section

- Apartment Rent—The 2012 amount is given. Use IF formulas to increase the rent by 10% if occupancy rates are high, or by 3% if occupancy rates are low.
- House Payments—This value is the total of the 12 monthly payments made on the mortgage. An important point to note is that house mortgage interest is always compounded *monthly*, not annually, as in the thrift shop tutorial. To properly calculate the house payments for the year, you divide the annual interest rate by 12 to determine the monthly interest. You also have to multiply a 30-year mortgage by 12 to get 360 payments, and then multiply the PMT formula by 12 to get the total amount for your annual house payments. Also, you will precede the PMT function with a negative sign to make the payment amount a positive number. Your formula should look like the following:

    =–PMT(B22/12,360,C5)*12
- Interest Rate for House Mortgage—Use the IF formula to enter a 3% interest rate if the bond money is available, and a 6% interest rate if no bond money is available.

Income and Cash Flow Statements Sections

As with the thrift shop tutorial, you want to see the Income and Cash Flow Statements for two scenarios—in this case, for continuing to rent and for purchasing a house. Each section begins with cash on hand at the end of 2012. As you can see in Figure C-36, Ted and Alice have only $4,000 in their savings.

	A	B	C	D
		2012	2013	2014
24	Income and Cash Flow Statement (Continue to Rent)	2012	2013	2014
25	Beginning-of-year Cash on Hand	NA		
26	Ted's Take Home Wages	$24,000		
27	Alice's Take Home Salary	$30,000		
28	Total Take Home Income	$54,000		
29	Apartment Rent	NA		
30	Utilities (Apartment)	NA		
31	Non-Housing Living Expenses	NA		
32	Total Expenses	NA		
33	End-of-year Cash on Hand	$4,000		
34				
35	Income and Cash Flow Statement (Purchase House)	2012	2013	2014
36	Beginning-of-year Cash on Hand	NA		
37	Ted's Take Home Wages	$24,000		
38	Alice's Take Home Salary	$30,000		
39	Total Take Home Income	$54,000		
40	House Payments	NA		
41	Real Estate Taxes and Insurance	NA		
42	Utilities (House)	NA		
43	House Repair and Maintenance Expense	NA		
44	Non-Housing Living Expenses	NA		
45	Total Expenses	NA		
46	End-of-year Cash on Hand	$4,000		

FIGURE C-36 Income and Cash Flow Statements sections (for both rent and purchase)

- Beginning-of-year Cash on Hand—This value is the End-of-year Cash on Hand from the previous year.
- Ted's Take Home Wages—This value is given for 2012. To get values for 2013 and 2014, increase Ted's wages by 2% each year.
- Alice's Take Home Salary—This value is given for 2012. To get values for 2013 and 2014, increase Alice's salary by 3% each year.
- Total Take Home Income—The sum of Ted and Alice's pay.
- Apartment Rent—The rent is copied from the Calculations section.
- House Payments—The house payments are also copied from the Calculations section.

- Real Estate Taxes and Insurance, Utilities (Apartment or House), House Repair and Maintenance Expense, and Non-Housing Living Expenses—These values all are copied from the Constants section.
- Total Expenses—This value is the sum of all the expenses listed above. Note that the house payment is now a positive number, so you can sum it normally with the other expenses.
- End-of-year Cash on Hand—This value is the Beginning-of-year Cash on Hand plus the Total Take Home Income minus the Total Expenses.

Scenario Manager Analysis

When you have completed the spreadsheet, set up Scenario Manager and create a Scenario Summary sheet. Ted and Alice want to look at their End-of-year Cash on Hand in 2014 for renting or buying under the following four scenarios:

- High occupancy and bond money available
- High occupancy and no bond money available
- Low occupancy and bond money available
- Low occupancy and no bond money available

If you have done your spreadsheet and Scenario Manager correctly, you should get the results shown in Figure C-37.

	A	B	C	D	E	F
1	Scenario Summary--Ted & Alice's House Purchase Decision--2014					
2			Hi Occ- Bond $	Hi Occ- No Bond $	Lo Occ- Bond $	Lo Occ- No Bond $
4	Changing Cells:					
5	Rental Occupancy (H-High, L-Low)	$C$12	H	H	L	L
6	Bond Mortgage Available (Y or N)	$C$13	Y	N	Y	N
7	Result Cells:					
8	End of Year Cash on Hand (Rent)	$D$16	$4,505	$4,505	$7,016	$7,016
9	End of Year Cash on Hand (Buy)	$D$17	$7,090	($1,452)	$7,090	($1,452)
10	Notes: Current Values column represents values of changing cells at					
11	time Scenario Summary Report was created. Changing cells for each					
12	scenario are highlighted in gray.					

FIGURE C-37   Scenario Summary results

Interpreting the Results

Based on the Scenario Summary results, what should Ted and Alice do? At first glance, it looks like the safe decision is to stay in the apartment. Actually, their decision hinges on whether they can get the lower-interest mortgage from the first-time buyers' bond issue. If they can, and if occupancy levels in apartments stay high, purchasing a house will give them about $2,500 more in savings at the end of 2014 than if they continued renting. Some other intangible factors are that home owners do not need permission to have pets, detached houses are quieter than apartments, and homes usually have a yard for pets and children to play in. Also, for the purposes of this exercise, you did not consider the tax benefits of home ownership. Depending on the amount of mortgage interest and real estate taxes Ted and Alice have to pay, they may be able to itemize their deductions and pay less income tax. If the income tax savings are more than $1,500, they can purchase the house even at the higher interest rate. In any case, because you did the DSS model for them, Ted and Alice now have a quantitative basis to help them make a good decision.

Visual Impact: Charting the Results

Charts and graphs often add visual impact to a Scenario Summary. Using the data from the Scenario Summary output table, try to create a chart similar to the one in Figure C-38 to illustrate the financial impact of each outcome.

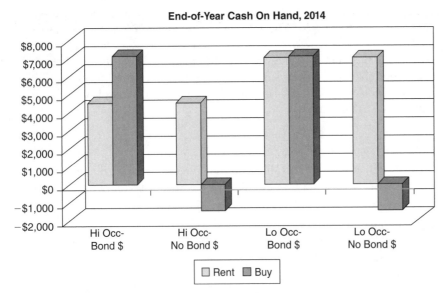

FIGURE C-38   A 3-D clustered column chart created from Scenario Summary data

Printing and Submitting Your Work

Ask your instructor which worksheets need to be printed for submission. Make sure your printouts of the spreadsheet, the Scenario Manager Summary table, and the graph (if you created one) fit on one printed page apiece. Click the File tab on the Ribbon, click the Print button, and then click Page Setup at the bottom of the Print Navigation pane. When the Page Setup dialog box opens, click the Page tab if it is not already open, then click the Fit to radio button and click 1 page wide by 1 page tall. Your spreadsheet, table, and graph will be fitted to print on one page apiece.

REVIEW OF EXCEL BASICS

This section reviews some basic operations in Excel and provides some tips for good work practices. Then you will work through some cash flow calculations. Working through this section will help you complete the spreadsheet cases in the following chapters.

Save Your Work Often—and in More Than One Place

To guard against data loss in case of power outages, computer crashes, and hard drive failure, it is always a good idea to save your work to a separate storage device. Copying a file into two separate folders on the same hard disk is *not* an adequate safeguard. If you are working on your college's computer network and you have been assigned network storage, the network storage is usually "mirrored"; in other words, it has duplicate drives recording data to prevent data loss if the system goes down. However, most laptops and home computers lack this feature. An excellent way to protect your work from accidental deletion is to purchase a USB "thumb" drive and copy all of your files to it.

When you save your Excel files, Windows will usually store them in the My Documents folder unless you specify the storage location. Instead of just clicking the Save icon, a good idea is to click the File group in the Ribbon (or the Office Button in Office 2007), and then click the Save As button. A dialog box will appear with icons on the left side, as shown in Figure C-39. If you have previously saved your file to a particular location, it will appear in the Save in text box at the top of the dialog box. To save the file in the same location, click Save. If your work is stored elsewhere, you can find the location using the icons on the left side of the dialog box. If you are saving to a USB thumb drive, it will appear as a storage device when you click the My Computer icon. Click the folder where you want to save your file.

NOTE

If you are trying an operation that might damage your spreadsheet and you do not want to use the Undo command, you can use the Save As command, and then add a number or letter to the filename to save an additional copy to "play with." Your original work will be preserved.

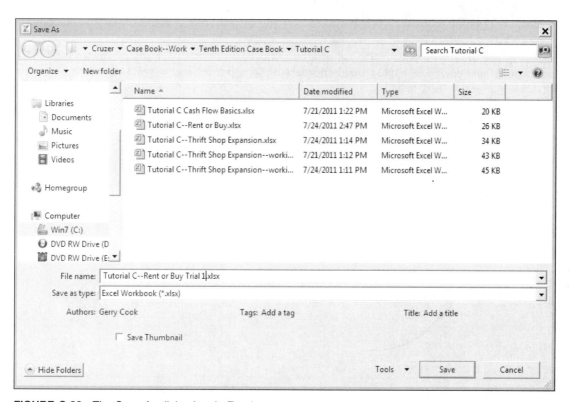

FIGURE C-39   The Save As dialog box in Excel

Basic Operations

To begin, you will review the following topics: formatting cells, displaying the spreadsheet cell formulas, circular reference errors, using the AND and OR logical operators in IF statements, and using nested IF statements to produce more than two outcomes.

Formatting Cells

Cell Alignment

Headings for columns are usually centered, while numbers in cells are usually aligned to the right. To set the alignment of cell data:

1.   Highlight the cell or cell range to format.
2.   Select the Home tab.
3.   In the Alignment group, click the button representing the horizontal alignment you want for the cell (Left Align, Center, or Right Align).
4.   Also in the Alignment group, above the horizontal alignment buttons, click the vertical alignment you want (Top Align, Middle, or Bottom Align). Middle Align is the most common vertical alignment for cells.

Cell Borders

Bottom borders are common for headings, and accountants include borders and double borders to indicate subtotals and grand totals on spreadsheets. Sometimes it is also useful to put a "box" border around a table of values or a section of a spreadsheet. To create borders:

1. Highlight the cell or cell range that needs a border.
2. Select the Home tab.
3. In the Font group, click the drop-down arrow of the Border icon. A menu of border selections appears (see Figure C-40).
4. Choose the desired border for the cell or group of cells. Note that All Borders creates a box border around each cell, while Outside Borders draws a box around a group of cells.

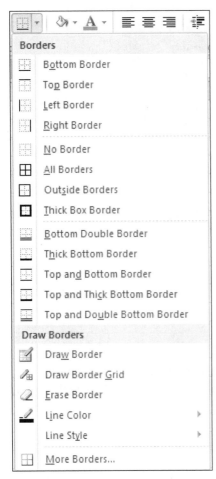

FIGURE C-40   Selections in the Borders menu

Number Formats

For financial numbers, you usually use the Currency format. (Do not use the Accounting format, as it places the $ sign to the far left side of the cell). To apply the appropriate Currency format:

1. Highlight the cell or cell range to be formatted.
2. Select the Home tab.
3. In the Number group, select Currency in the Number Format drop-down list.
4. To set the desired number of decimal places, click the Increase Decimal or Decrease Decimal button in the bottom-right corner of the group (see Figure C-41).

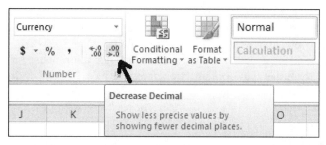

FIGURE C-41 Increase Decimal and Decrease Decimal buttons

If you do not know what a button does in Office, hover your mouse pointer over the button to see a description.

Format "Painting"

If you want to copy *all* the format properties of a certain cell to other cells, use the Format Painter. First, select the cell whose format you want to copy. Then click the Format Painter button (the paintbrush icon) in the Clipboard Group under the Home tab (see Figure C-42). When you click the button, the mouse pointer turns into a paintbrush. Click the cell you want to reformat. To format multiple cells, select the cell whose format you want to copy, and then click *twice* on the Format Painter button. The mouse cursor will become a paintbrush, and the paint function will stay on so you can reformat as many cells as you want. To turn off the Format Painter, click its button again or press the Esc key.

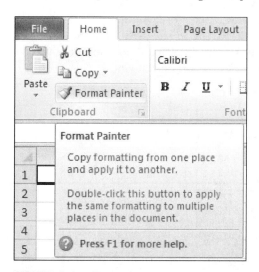

FIGURE C-42 The Format Painter button

Showing the Excel Formulas in the Cells

Sometimes your instructor might want you to display or print the formulas in the spreadsheet cells. If you want the spreadsheet cells to display the actual cell formulas, follow these steps:

1. While holding down the Ctrl key, press the key in the upper-left corner of the keyboard that contains the back quote (`) and tilde (~). The spreadsheet will display the formulas in the cells. The columns may also become quite wide—if so, do not resize them.
2. The Ctrl+`~ key combination is a toggle; to restore your spreadsheet to the normal cell contents, press Ctrl+`~ again.

Understanding Circular Reference Errors

When entering formulas, you might make the mistake of referring to the cell in which you are entering the formula as part of the formula, even though it should only display the output of that formula. Referring a cell back to itself in a formula is called a *circular reference*. For example, suppose that in cell B2 of a worksheet, you enter =B2-B1. A terrible but apt analogy for a circular reference is a cannibal trying to eat himself! Fortunately, Excel informs you when you try to enter a circular reference into a formula (see Figure C-43). Excel also warns you if you try to open an existing spreadsheet that has one or more circular references. Before you can use the spreadsheet, you must fix the formulas that contain circular references.

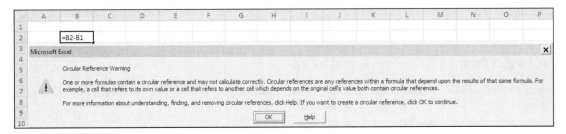

FIGURE C-43   Excel circular reference warning

Using AND and OR Functions in IF Statements

Recall that the IF function has the following syntax:

> IF(test condition, result if test is True, result if test is False)

The test conditions in the previous example IF statements tested only one cell's value, but a test condition can test more than one value of a cell.

For example, look at the thrift shop tutorial again. The Total Sales Dollars for 2013 depended on the economic outlook (recession or boom). The original IF statement was =IF($C$8="R",B18*1.3,B18*1.15), as shown in Figure C-44. This function increased the 2012 Total Sales Dollars by 30% if there was continued recession ("R" entered in cell C8), but only increased the total by 15% if there was a boom.

C18	f_x =IF($C$8="R",B18*1.3,B18*1.15)		
A	B	C	D
7 Inputs	**2012**	**2013**	**2014**
8 Economic Outlook (R=Recession, B=Boom)	NA	R	NA
9 Inflation Outlook (H=High, L=Low)	NA	H	NA
10			
11 Summary of Key Results	**2012**	**2013**	**2014**
12 Net Income after Taxes (Expansion)	NA	$69,974	$73,660
13 End-of-year Cash on Hand (Expansion)	NA	$84,974	$158,634
14 Net Income after Taxes (No Expansion)	NA	$72,601	$69,936
15 End-of-year Cash on Hand (No Expansion)	NA	$87,601	$157,537
16			
17 Calculations (Expansion)	**2012**	**2013**	**2014**
18 Total Sales Dollars	$350,000	$455,000	$591,500
19 Cost of Goods Sold	$245,000	$337,810	$465,227
20 Cost of Goods Sold (as a percent of Sales)	70%	74%	79%
21 Interest Rate for Business Loan	5%	NA	NA

FIGURE C-44   The original IF statement used to calculate the Total Sales Dollars for 2013

To take the IF argument one step further, assume that the Total Sales Dollars for 2013 depended not only on the Economic Outlook, but on the Inflation Outlook (High or Low). Suppose there are two possibilities:

- Possibility 1: If the economic outlook is for a Recession and the inflation outlook is High, the Total Sales Dollars for 2013 will be 30% higher than in 2012.
- Possibility 2: For the other three cases (Recession and Low Inflation, Boom and High Inflation, and Boom and Low Inflation), assume that the 2013 Total Sales Dollars will only be 15% higher than in 2012.

The first possibility requires two conditions to be true at the same time: C8="R" and C9="H". You can include an AND( ) function inside the IF statement to reflect the additional condition as follows:

=IF(AND($C$8="R",$C$9="H"), B18*130%,B18*115%)

When the test argument uses the AND() function, conditions "R" *and* "H" both must be present at the same time for the statement to use the true result (multiplying last year's sales by 130%). Any of the other three outcome combinations will cause the statement to use the false result (multiplying last year's sales by 115%).

You can also use an OR() function in an IF statement. For example, assume that instead of both conditions (Recession and High Inflation) having to be present, only one of the two conditions needs to be present for sales to increase by 30%. In this case, you use the OR() function in the test argument as follows:

=IF(OR($C$8="R",$C$9="H"), B18*130%,B18*115%)

In this case, if *either* of the two conditions (C8="R" or C9="H"), is true, the function will return the true argument, multiplying the 2012 sales by 130%. If *neither* of the two conditions is true, then the function will return the false argument, multiplying the 2012 sales by 115% instead.

Using IF Statements Inside IF Statements (Also Called "Nesting IFs")

By now you should be familiar with IF statements, but here is a quick review of the syntax:

=IF(test condition, result if test is True, result if test is False)

In the preceding examples, only two courses of action were possible for each of the inputs: Recession or Boom, High Inflation or Low Inflation, Rental Occupancy High or Low, Bond Money Available or No Bond Money Available. The tutorial used only two possible outcomes to keep them simple.

However, in the business world, decision support models are frequently based on three or more possible outcomes. For capital projects and new product launches, you will frequently project financial outcomes based on three possible scenarios: Most Likely, Worst Case, and Best Case. You can modify the IF statement by placing another IF statement inside the result argument if the first test is false, creating the ability to launch two more alternatives from the second IF statement. This is called "nesting" your IF statements.

Try a simple nested IF statement: In your thrift shop example, assume that three economic outlooks are possible: Recession (R), Boom (B), or Stable (S). As before, the 2013 Total Sales Dollars (cell C18) will be the 2012 Total Sales Dollars increased by some fixed percentage. In a Recession, sales will increase by 30%, in a Boom they will increase by 15%, and for a Stable Economic Outlook, sales will increase by 22%, which is roughly midway between the other two percentages. You can "nest" the IF statement in cell C18 to reflect the third outcome as follows:

=IF($C$8= "R",B18*130%,IF($C$8="B",B18*115%,B18*122%))

Note the added IF statement inside the False value argument. You can break down this statement:

- If the value in cell C8 is "R", multiply the value in cell B18 by 130%, and enter the result in cell C18.
- If the value in cell C8 is not "R", check whether the value in cell C8 is "B". If it is, multiply the value in cell B18 by 115%, and enter the result in cell C18.
- If the value in cell C8 is not "B", multiply the value in cell B18 by 122%, and enter the result in cell C18.

If you have four or more alternatives, you can keep nesting IF statements inside the false argument for the outer IF statements. (Excel 2007 and later versions have a limit of 64 levels of nesting in the IF function, which should take care of every conceivable situation.)

NOTE

The "embedded IFs" in a nested IF statement are not preceded by an equals sign. Only the first IF gets the equals sign.

Cash Flow Calculations: Borrowing and Repayments

The Scenario Manager cases that follow in this book require accounting for money that the fictional company will have to borrow or repay. This money is not like the long-term loan that the Collegetown Thrift Shop is considering for its expansion. Instead, this money is short-term borrowing that companies use to pay current obligations, such as purchasing inventory or raw materials. Such short-term borrowing is called a line of credit, and is extended to businesses by banks, much like consumers have credit cards. Lines of credit usually involve interest payments, but for simplicity's sake, focus instead on how to do short-term borrowing and repayment calculations.

To work through cash flow calculations, you must make two assumptions about a company's borrowing and repayment of short-term debt. First, you assume that the company has a desired *minimum* cash level at the end of a fiscal year (which is also its cash level at the start of the next fiscal year), to ensure that the company can cover short-term expenses and purchases. Second, assume the bank that serves the company will provide short-term loans (a line of credit) to make up the shortfall if the end-of-year cash falls below the desired minimum level.

NCP stands for Net Cash Position, which equals beginning-of-year cash plus net income after taxes for the year. NCP represents the available cash at the end of the year, *before* any borrowing or repayment. For the three examples shown in Figure C-45, set up a simple spreadsheet in Excel and determine how much the company needs to borrow to reach its minimum year-end cash level. Use the IF function to enter 0 under Amount to Borrow if the company does not need to borrow any money.

	A	B	C	D
			Minimum Cash	Amount to
1	Example	NCP	Required	Borrow
2	1	$25,000	$10,000	?
3	2	$9,000	$10,000	?
4	3	($12,000)	$10,000	?

FIGURE C-45   Examples of borrowing

You can also assume that the company will use some of its cash on hand at the end of the year to pay off as much of its outstanding debt as possible without going below its minimum cash on hand required. The "excess" cash is the company's NCP *less* the minimum cash on hand required—any cash above the minimum is available to repay any debt. In the examples shown in Figure C-46, compute the excess cash and then compute the amount to repay. In addition, compute the ending cash on hand after the debt repayment.

	A	B	C	D	E	F
			Minimum Cash	Beginning-of-Year		Ending
9	Example	NCP	Required	Debt	Repay?	Cash
10	1	$12,000	$10,000	$5,000	?	?
11	2	$13,000	$10,000	$8,000	?	?
12	3	$20,000	$10,000	$0	?	?
13	4	$60,000	$10,000	$40,000	?	?
14	5	($20,000)	$10,000	$10,000	?	?

FIGURE C-46   Examples of debt repayment

In the Scenario Manager cases of the following chapters, your spreadsheets may need two bank financing sections beneath the Income and Cash Flow Statements sections. You will build the first section to calculate any needed borrowing or repayment at year's end to compute year-end cash on hand. The second section will calculate the amount of debt owed at the end of the year after any borrowing or repayment.

Return to the Collegetown Thrift Shop tutorial and assume that it includes a line of credit at a local bank for short-term cash management. The first new section extends the end-of-year cash calculation, which was shown for the thrift shop in Figure C-19. Figure C-47 shows the structure of the new section highlighted in boldface.

A	B	C	D
29 Income and Cash Flow Statements (Expansion)	2012	2013	2014
30 Beginning-of-year Cash on Hand	NA	$15,000	$0
31 Sales (Revenue)	NA	$455,000	$591,500
32 Cost of Goods Sold	NA	$337,610	$465,227
33 *Business Loan Payment*	NA	($12,950)	($12,950)
34 Income before Taxes	NA	$104,440	$113,323
35 Income Tax Expense	NA	$34,465	$39,663
36 Net Income after Taxes	NA	$69,974	$73,660
Net Cash Position NCP			
Beginning-of-year Cash on Hand			
37 **plus Net Income after Taxes**	NA	$84,974	$73,660
38 **Line of credit borrowing from bank**	NA		
39 **Line of credit repayments to bank**	NA		
40 **End-of Year Cash on Hand**	**$15,000**		

FIGURE C-47   Calculation section for End-of-Year Cash on Hand with borrowing and repayments added

The heading in cell A36 was originally End-of-year Cash on Hand in Figure C-19, but you will add line-of-credit borrowing and repayment to the end-of-year totals. You must add the line-of-credit borrowing from the bank to the NCP and subtract the line-of-credit repayments to the bank from the NCP to obtain the End-of-Year Cash on Hand.

The second new section you add will compute the End-of year debt owed. This section is called Debt Owed, as shown in Figure C-48.

A	B	C	D
42 **Debt Owed**	2012	2013	2014
43 **Beginning-of-year debt owed**	NA		
44 **Borrowing from bank line of credit**	NA		
45 **Repayment to bank line of credit**	NA		
46 **End-of-year debt owed**	**$47,000**		

FIGURE C-48   Debt Owed section

As you can see, the thrift shop currently owes $47,000 on its line of credit at the end of 2012. The End-of-year debt owed equals the Beginning-of-year debt owed plus any new borrowing from the bank's line of credit, minus any repayment to the bank's line of credit. Therefore, the formula in cell C46 would be:

=C43+C44-C45

Assume that the amounts for borrowing and repayment (cells C44 and C45) were calculated in the first new section (for the year 2013, the amounts would be in cells C38 and C39), and then copied into the second section. The formula for cell C44 would be =C38, and for cell C45 would be =C39. The formula for cell C43, Beginning-of-year debt owed in 2013, would simply be the End-of-year debt owed in 2012, or =B46.

Now that you have added the spreadsheet entries for borrowing and repayment, consider the logic for the borrowing and repayment formulas.

Calculation of Borrowing from the Bank Line of Credit

When using logical statements, it is sometimes easier to state the logic in plain language and then turn it into an Excel formula. For borrowing, the logic in plain language is:

If (cash on hand before financing transactions is greater than the minimum cash required,

then borrowing is not needed; else,

borrow enough to get to the minimum)

You can restate this logic as the following:

If (NCP is greater than minimum cash required,

then borrowing from bank=0;

else, borrow enough to get to the minimum)

You have not added minimum cash at the end of the year as a requirement, but you could add it to the Constants section at the top of the spreadsheet (in this case the new entry would be cell C6). Assume that you want $50,000 as the minimum cash on hand at the end of both 2013 and 2014. Assuming that the NCP is shown in cell C37, you could restate the formula for borrowing (cell C38) as the following:

IF(NCP>Minimum Cash, 0; otherwise, borrow enough to get to the minimum cash)

You have cell addresses for NCP (cell C37) and for Minimum Cash (cell $C$6). To develop the formula for cell C38, substitute the cell address for the test argument; the true argument is simply zero (0), and the false argument is the minimum cash minus the current NCP. The formula stated in Excel for cell C38 would be:

=IF(C37>=$C$6, 0, $C$6-C37)

Calculation of Repayment to the Bank Line of Credit

Simplify the statements first in plain language:

IF(beginning of year debt=0, repay 0 because nothing is owed, but

IF(NCP is less than the minimum, repay 0, because you must borrow, but

IF(extra cash equals or exceeds the debt, repay the whole debt,

ELSE (to stay above the minimum cash, repay the extra cash above the minimum)

Look at the following formula. If you assume that the repayment amount will be in cell C39, the beginning-of-year debt is in cell C43, and the minimum cash target is still in cell $C$6, the repayment formula for cell C39 with the nested IFs should look like the following:

=IF(C43=0,0,IF(C37<=$C$6,0,IF(C37-$C$6>=C43,C43,C37-$C$6)))

The new sections of the thrift shop spreadsheet would look like those in Figure C-49.

	A	B	C	D
29	**Income and Cash Flow Statements (Expansion)**	**2012**	**2013**	**2014**
30	Beginning-of-year Cash on Hand	NA	$15,000	$50,000
31	Sales (Revenue)	NA	$455,000	$591,500
32	Cost of Goods Sold	NA	$337,610	$465,227
33	*Business Loan Payment*	NA	($12,950)	($12,950)
34	Income before Taxes	NA	$104,440	$113,323
35	Income Tax Expense	NA	$34,465	$39,663
36	Net Income after Taxes	NA	$69,974	$73,660
	Net Cash Position NCP			
	Beginning-of-year Cash on Hand			
37	**plus Net Income after Taxes**	NA	$84,974	$123,660
38	**Line of credit borrowing from bank**	NA	$0	$0
39	**Line of credit repayments to bank**	NA	$34,974	$12,026
40	**End-of Year Cash on Hand**	$15,000	$50,000	$111,634
41				
42	**Debt Owed**	**2012**	**2013**	**2014**
43	**Beginning-of-year debt owed**	NA	$47,000	$12,026
44	**Borrowing from bank line of credit**	NA	$0	$0
45	**Repayment to bank line of credit**	NA	$34,974	$12,026
46	**End-of-year debt owed**	$47,000	$12,026	$0

FIGURE C-49 Thrift shop spreadsheet with line-of-credit borrowing, repayments, and Debt Owed added

Answers to the Questions about Borrowing and Repayment

Figures C-50 and C-51 display solutions for the borrowing and repayment calculations.

	A	B	C	D
1	Example	NCP	Minimum Cash Required	Amount to Borrow
2	1	$25,000	$10,000	$0
3	2	$9,000	$10,000	$1,000
4	3	($12,000)	$10,000	$22,000

FIGURE C-50   Answers to examples of borrowing

In Figure C-50, the formula in cell D2 for the amount to borrow is =IF(B2>=C2,0,C2-B2).

	A	B	C	D	E	F
9	Example	NCP	Minimum Cash Required	Beginning-of-Year Debt	Repay?	Ending Cash
10	1	$12,000	$10,000	$5,000	$2,000	$10,000
11	2	$13,000	$10,000	$8,000	$3,000	$10,000
12	3	$20,000	$10,000	$0	$0	$20,000
13	4	$60,000	$10,000	$40,000	$40,000	$20,000
14	5	($20,000)	$10,000	$10,000	$0	NA

FIGURE C-51   Answers to examples of repayment

In Figure C-51, the formula in cell E10 for the amount to repay is

=IF(B10>=C10,IF(D10>0,MIN(B10-C10,D10),0),0).

Note the following points about the repayment calculations shown in Figure C-51.

- In Example 1, only $2,000 is available for debt repayment ($12,000 – $10,000) to avoid dropping below the Minimum Cash Required.
- In Example 2, only $3,000 is available for debt repayment.
- In Example 3, the Beginning-of-Year Debt was zero, so the Ending Cash is the same as the Net Cash Position.
- In Example 4, there was enough cash to repay the entire $40,000 debt, leaving $20,000 in Ending Cash.
- In Example 5, the company has cash problems—it cannot repay any of the Beginning-of-Year Debt of $10,000, and it will have to borrow an additional $30,000 to reach the Minimum Cash Required target of $10,000.

You should now have all the basic tools you need to tackle Scenario Manager in Cases 6 and 7. Good luck!

CASE **6**

BUFFALO BEN'S GAME RANCH EXPANSION DECISION

Decision Support Using Excel

PREVIEW

Buffalo Ben's Game Ranch is a privately owned company in central Wyoming. Ben Johnson, the owner, started the ranch in the 1990s when demand was increasing for domestically raised buffalo. Bison meat, according to the USDA Nutritional Database, contains less fat and cholesterol than beef, pork, or chicken.

Buffalo Ben's has grown in the past two decades. Recently, Ben acquired several hundred acres of range adjacent to his ranch. Originally he had planned to expand his buffalo production on the new land. However, two drought years in the past decade resulted in net income losses, primarily because the buffalo had to be fed grain when the range grass dried up. Ben is thinking of diversifying his game ranch by raising ostriches on the new acreage. Ostriches do not require as much additional feed as buffalo during a drought, and ostrich meat, eggs, and feathers are in high demand. Ostrich meat also ranks highly in the USDA Nutritional Database, equaling buffalo in almost every nutritional category.

Ben has done some initial research on ostrich farming, and has good estimates of the investment and operating costs required to convert the new acreage to an ostrich range. He also traveled to Texas to recruit potential ostrich wranglers. While the possibility of bringing ostrich ranching to Wyoming excites Ben, he realizes that it is a potentially risky venture.

Ben has hired you to do a comparative financial analysis of the choice between expanding the buffalo ranch or creating an ostrich ranch on the new acreage. Ben's accountant and ranch manager have developed some initial inputs for you to create an investment decision model using an Excel workbook. Your finished model will be instrumental in helping Buffalo Ben's choose a wise expansion strategy.

PREPARATION

- Review the spreadsheet concepts discussed in class and in your textbook.
- Your instructor may assign Excel exercises to help prepare you for this case.
- Tutorial C has an excellent review of IF, nested IF, and AND statements that will help you with this case.
- Review the file-saving instructions in Tutorial C—it is always a good idea to save an extra copy of your work on a USB thumb drive.
- Reviewing Tutorial F will help you brush up on your presentation skills.
- Because Buffalo Ben's case is a strategic investment decision model, you will calculate the internal rate of return (IRR). If you are unfamiliar with the IRR function in Excel, this case includes a section that explains how to set it up.

BACKGROUND

You are an information analyst for Buffalo Ben's. Ben has asked you to prepare a quantitative analysis of financial, sales, and operations data to help determine which expansion path would offer the best strategic opportunity for the ranch. Ben's managers have been asked to provide data from their respective functional areas:

- Accounting—The current cash position of the company, the cash outlay for the two investment choices, data for feed, labor, and operating costs for each alternative, and the corporate income tax rates
- Sales—Forecasts for the wholesale (meat processing) price of buffalo and ostrich based on the current national and global economy and demand
- Ranch operations—Labor costs for buffalo and ostrich wranglers, feed consumption, and overhead costs, including utilities, fuel, horses, and veterinary costs

Specifically, the departments have given you the following data:

- Feed costs per day per buffalo or ostrich, 2013 through 2015
- Direct labor costs per day for buffalo or ostrich wranglers, 2013 through 2015
- Overhead expenses per day for the ranch expansion, 2013 through 2015
- Stable market price of buffalo meat (on the hoof) and ostrich meat, 2013 through 2015
- Capital investment dollars for buffalo or ostrich ranch expansion (end of 2012)
- Buffalo or ostrich capacity per acre
- Total employment days in the ranching season
- Additional acres of land purchased
- Corporate income tax rates, 2013 through 2015

In addition, the Assignment 1 section contains information you need to write the formulas for the Calculations section, Income and Cash Flow Statements section, and IRR Calculation section.

You will use Excel to project how much profit and positive cash flow each expansion alternative will generate for Buffalo Ben's for the next three years. You also will use Excel to calculate an internal rate of return for each alternative. Finally, you will examine the effects of the economy and the effects of rainfall on your projected sales and profits for each alternative. In summary, your DSS will include the following inputs:

1. Your decision to invest in either the buffalo or ostrich ranch expansion
2. Whether the economic outlook is for a recession or stable cycle
3. Whether the rainfall is normal or indicates a drought

In a stable economy, the demand for exotic meat is greater than the available world supply. However, a recession would drive down demand and prices for bison and ostrich meat slightly.

One factor that is difficult for Buffalo Ben's to quantify is its lack of experience in ostrich ranching. Abundant data exists from other ostrich ranches (mostly in Texas), but Ben realizes that introducing ostriches into the ranch will bring unexpected challenges and risks. He is not sure whether he has anticipated those risks, but he knows that the basis for any good decision is a well-reasoned financial analysis.

Your DSS model must account for the effects of the preceding three inputs on costs, selling prices, and other variables. If you design the model well, it will let you develop "what-if" scenarios with all the inputs, see the results, and show a preferred alternative for Buffalo Ben's Game Ranch to adopt.

ASSIGNMENT 1: CREATING A SPREADSHEET FOR DECISION SUPPORT

In this assignment, you will create a spreadsheet that models the business decision Buffalo Ben's is seeking. In Assignment 2, you'll write a report to Ben about your analysis and recommendations. In Assignment 3, you will prepare and give a presentation of your analysis and recommendations.

You will start by creating the spreadsheet model of the company's financial and operating data. The model will cover three years of ranching operations and sales (2013 through 2015) for the alternative ranch expansions considered. Assume that Ben and his accountant have completed the preliminary research in 2012, and that the new acreage, associated buildings, and equipment are in place to begin operations and sales in 2013.

This section will help you set up each of the following spreadsheet components before entering the cell formulas:

- Constants
- Inputs
- Summary of Key Results

- Calculations
- Income and Cash Flow Statements
- Internal Rate of Return Calculation

The internal rate of return was not calculated in the Tutorial C examples. The section was added to this case because Excel financial formulas such as IRR work better if the cash outflow and inflow data is arranged in a vertical column with the years in ascending order, as opposed to taking the cash flows from across the page or from nonadjacent cells.

The spreadsheet skeleton is available for you to use. To access this skeleton, go to your data files, select Case 6, and then select **Case 6—Buffalo Ben Game Ranch.xlsx**.

Constants Section

First, build the skeleton of your spreadsheet. Set up your Constants section as shown in Figure 6-1. An explanation of the line items follows the figure.

	A	B	C	D	E
1	**Buffalo Ben's Game Ranch Expansion**				
2					
3	**Constants**	**2012**	**2013**	**2014**	**2015**
4	Feed Cost per day per Buffalo (normal rainfall)	NA	$11.00	$11.33	$11.67
5	Feed Cost per day per Ostrich (normal rainfall)	NA	$3.75	$3.86	$3.98
6	Direct Labor Cost/Day, additional Buffalo Wranglers	NA	$1,600	$1,680	$1,764
7	Direct Labor Cost/Day, Ostrich Wranglers	NA	$2,000	$2,100	$2,205
8	Overhead Expenses per day for ranch expansion		$300	$315	$331
9	Stable Market Price of Buffalo Meat (on the Hoof, Yearling Calf)	NA	$3,000	$3,090	$3,183
10	Stable Market Price of Ostrich Meat and Feathers (Young Adult)	NA	$1,100	$1,133	$1,167
11	Capital Investment for Buffalo Ranch Expansion	$1,000,000	NA	NA	NA
12	Capital Investment for Ostrich Ranch Addition	$2,000,000	NA	NA	NA
13	Buffalo Capacity per Acre (grazing)	2	NA	NA	NA
14	Ostrich Capacity per Acre (grazing and running)	7	NA	NA	NA
15	Total Employment Days--Ranching Season	200	NA	NA	NA
16	Additional Acres of Land Purchased	800	NA	NA	NA
17	Corporate Income Tax Rate	25%	25%	26%	27%

FIGURE 6-1 Constants section

- Feed Cost per day per Buffalo (normal rainfall)—This value is the daily cost of grain fed to one buffalo. The daily feed cost will increase significantly in a drought; this possibility is addressed in the Calculations section.
- Feed Cost per day per Ostrich (normal rainfall)—This value is the cost of feeding one ostrich per day. The daily feed costs for ostriches will increase slightly in a drought; this possibility is addressed in the Calculations section.
- Direct Labor Cost/Day, additional Buffalo Wranglers—This value is the cost per day for buffalo wranglers if the company expands the buffalo ranch.
- Direct Labor Cost/Day, Ostrich Wranglers—This value is the cost per day for ostrich wranglers hired if the company chooses ostriches for expansion.
- Overhead Expenses per day for ranch expansion—This value is the overhead expense per day for the ranch expansion. Overhead includes fuel, maintenance, utilities, and veterinary costs. Ben has estimated the expenses to be the same for either buffalo or ostriches.
- Stable Market Price of Buffalo Meat (on the Hoof, Yearling Calf)—This value estimates the market price for one live buffalo calf sold for processing. The value assumes a stable economy. In a recession, the market price will decrease.
- Stable Market Price of Ostrich Meat and Feathers (Young Adult)—This value estimates the market price for one live adult ostrich sold for processing. The value assumes a stable economy.
- Capital Investment for Buffalo Ranch Expansion—This value is the total amount of investment money needed to complete the expansion for buffalo ranching and furnish it with breeding stock.
- Capital Investment for Ostrich Ranch Addition—This value is the total amount of investment money required to complete the addition for ostrich ranching and furnish it with breeding stock.

- Buffalo Capacity per Acre (grazing)—This value is the number of buffalo that each acre of the expansion can accommodate for grazing.
- Ostrich Capacity per Acre (grazing and running)—This value is the number of ostriches that each acre of the expansion can accommodate.
- Total Employment Days—Ranching Season—This value estimates the number of days per year that the company must employ wranglers after expanding.
- Additional Acres of Land Purchased—This value is the number of acres that Ben purchased for expanding the ranch.
- Corporate Income Tax Rate—This value estimates the corporate income tax rates that the ranch will have to pay after the expansion.

Inputs Section

Your spreadsheet model must include the following inputs, which will apply for all three years (see Figure 6-2).

A	B	C	D	E
19 Inputs	2012	2013	2014	2015
20 Annual Rain (N-Normal, D-Drought)	NA		NA	NA
21 Economic Outlook (S-Stable, R-Recession)	NA		NA	NA
22 Expansion Selection (B-Buffalo, O-Ostrich)	NA		NA	NA

FIGURE 6-2 Inputs section

- Annual Rain—This value is Normal (N) or Drought (D). A drought will reduce the amount of grass available for grazing and require the game ranch to spend more money on feed for buffalo or ostriches.
- Economic Outlook—This value is Stable (S) or Recession (R). The economic outlook affects the demand for exotic meat. For example, the ranch will get a lower price for its meat during a recession.
- Expansion Selection—This value is the basic input for the strategic decision to expand the ranch to raise more buffalo (B) or to start raising ostriches (O).

Summary of Key Results Section

This section (see Figure 6-3) contains the results data, which is of primary interest to the management team of Buffalo Ben's Game Ranch. The data includes information for income and end-of-year cash on hand, as well as the internal rate of return (annualized) for a particular set of business inputs. This section summarizes the values from the Calculations, Income and Cash Flow Statements, and Internal Rate of Return Calculation sections.

A	B	C	D	E
24 Summary of Key Results	2012	2013	2014	2015
25 Net Expansion Income after Taxes	NA			
26 End-of-year Cash on hand	NA			
27 Internal Rate of Return for Investment	NA	NA	NA	

FIGURE 6-3 Summary of Key Results section

For each year from 2013 to 2015, your spreadsheet should show net income after taxes (which will also be the cash inflows for the IRR calculation) and end-of-year cash on hand. Because Ben is funding the capital investment from his own cash on hand at the end of 2012, there is no debt to repay. However, Ben wants to know the IRR at the end of 2015.

Calculations Section

The Calculations section includes the calculations needed to determine the total costs of raising animals for sale and the number of animals raised for sale. See Figure 6-4.

	A	B	C	D	E
29	**Calculations**	**2012**	**2013**	**2014**	**2015**
30	New Product Capital Investment (internally financed)		NA	NA	NA
31	Total Feed Costs	NA			
32	Total Direct Labor	NA			
33	Total Expansion Overhead	NA			
34	Cost of Goods Sold	NA			
35	Total Animals raised for sale	NA			

FIGURE 6-4 Calculations section

- New Product Capital Investment—This is the amount of investment money that Ben spent at the end of 2012, depending on the Expansion Selection from the Inputs section. If Buffalo (B) is selected, this cell should display the capital investment amount from the Constants section for buffalo (the value in cell B11). If Ostrich (O) is selected, the cell should display the capital investment amount from the Constants section for ostriches (the value in cell B12). Note that when you see the word "if" in the text, you need to write a formula using the IF function for the target cell.

- Total Feed Costs—You are given the daily feed costs (assuming normal rainfall) for buffalo and ostriches in the Constants section. You must write formulas to calculate the Total Feed Costs for 2013, 2014, and 2015. The basic formula is:

  Number of animals raised per acre × Total acres in expansion × Number of ranching days per year × Feed cost per day per animal

  All of these formula values are shown in the Constants section. Ben's team members also believe that the preceding formula will have to be modified based on the expansion selection and whether a drought occurs. There are four possible combinations and results:

  - Buffalo and Normal Rainfall—For this scenario, the total feed costs will be calculated using the preceding formula. (Remember to use the buffalo feed costs.)
  - Buffalo and Drought—The total feed costs will be calculated using the basic formula multiplied by 1.15 to reflect 15 percent more feed consumption by buffalo because less grass is available for grazing.
  - Ostriches and Normal Rainfall—The total feed costs will be calculated using the preceding formula. (Remember to use the ostrich feed costs.)
  - Ostriches and Drought—The total feed costs will be calculated using the basic formula multiplied by 1.05 to reflect 5 percent more feed consumption by ostriches because less grass is available for grazing.
    (*Hint*: For these formulas, you must use a nested IF statement that contains AND functions.)

- Total Direct Labor—Using the daily direct labor costs for buffalo wranglers and ostrich wranglers from the Constants section, develop the basic formula for Total Direct Labor cost: the direct labor cost per day for buffalo wranglers or ostrich wranglers multiplied by the Total Employment Days—Ranching Season. Write the formula as an IF statement to calculate the Total Direct Labor for either the buffalo or ostrich expansion.

- Total Expansion Overhead—This value is the overhead expenses per day multiplied by the Total Employment Days—Ranching Season. Both factors are shown in the Constants section.

- Cost of Goods Sold—This is the sum of the Total Feed Costs, Total Direct Labor, and Total Expansion Overhead.

- Total Animals raised for sale—If the buffalo expansion is selected, this value is the Buffalo Capacity per Acre multiplied by the Additional Acres of Land Purchased. If the ostrich expansion is selected, the value is the Ostrich Capacity per Acre multiplied by the Additional Acres of Land Purchased. These factors are shown in the Constants section.

The Calculations section includes several complicated formulas. It is easiest to write all the formulas for one column (Year 2013), and then copy the formulas to the other two year columns. If you do this, be sure to use *absolute* cell references for any constants you use from cells B13 through B16 (that is, $B$13, $B$14, $B$15, and $B$16); otherwise, your calculations will not copy correctly to the next two years. If you need help writing the nested IF and AND formulas, refer to Tutorial C or ask your instructor.

Income and Cash Flow Statements Section

The statements for income and cash flow start with the cash on hand at the beginning of the year. Because Buffalo Ben's is funding the capital investment *internally*—that is, with its own cash on hand—you must deduct the invested funds from the cash on hand at the end of year 2012. Figure 6-5 and the following list show you how to structure the Income and Cash Flow Statements section.

	A	B	C	D	E
37	**Income and Cash Flow Statements**	**2012**	**2013**	**2014**	**2015**
38	Beginning-of-year Cash on Hand (in 2013 deduct Investment for 2012)	NA			
39	Ranch Expansion Sales Revenues	NA			
40	less: Cost of Goods Sold	NA			
41	Expansion Profit before Income Tax	NA			
42	less: Income Tax Expense	NA			
43	Net Expansion Income after Taxes (Cash Inflow)	NA			
44	End-of-year Cash on Hand	$2,500,000			

FIGURE 6-5 Income and Cash Flow Statements section

- Beginning-of-year Cash on Hand—For 2013, this value is the End-of-year Cash on Hand from 2012 minus the capital investment, depending on the expansion selection. If you choose the buffalo expansion, the capital investment will be $1 million (cell B11 in the Constants section). If you choose the ostrich expansion, the capital investment will be $2 million (cell B12 in the Constants section). For 2014 and 2015, the Beginning-of-year Cash on Hand will be the End-of-year Cash on Hand from the previous year.
- Ranch Expansion Sales Revenues—This value depends on the selected expansion (buffalo or ostrich) and the state of the economy (stable or recession). Using 2013 as an example, the basic formula for sales revenues is the market price per animal (cell C9 or C10, depending on the expansion selection) multiplied by the Total Animals raised for sale (cell C35). There are four possible combinations and results:

  - Buffalo and Stable—The sales revenues will be calculated using the preceding formula. (Remember to use the buffalo market price.)
  - Buffalo and Recession—The sales revenues will be calculated using the basic formula multiplied by .95 to reflect a 5 percent reduction in market price for buffalo because of less demand in a weaker economy.
  - Ostriches and Stable—The sales revenues will be calculated using the preceding formula. (Remember to use the ostrich market price.)
  - Ostriches and Recession—The sales revenues will be calculated using the basic formula multiplied by .95 to reflect a 5 percent reduction in market price for ostriches because of less demand in a weaker economy.

    (*Hint*: For these formulas, you must use a nested IF statement that contains AND functions.)
- Less: Cost of Goods Sold—This value is the Cost of Goods Sold copied from the Calculations section.
- Expansion Profit before Income Tax—This value is the Ranch Expansion Sales Revenues minus the Cost of Goods Sold.
- Less: Income Tax Expense—If you make a profit (in other words, if the Expansion Profit before Income Tax is greater than zero), this value is the Expansion Profit before Income Tax multiplied by the Corporate Income Tax Rate from the Constants section. If you make nothing or have a net loss, the Income Tax Expense is zero.
- Net Expansion Income after Taxes (Cash Inflow)—This value is the Expansion Profit before Income Tax minus the Income Tax Expense. From a strict accounting standpoint, the Net Expansion Income after Taxes is not the cash inflow, because you would have to add all noncash expenses such as depreciation and/or depletion to determine true cash inflow. However, for the purposes of this case, assume that Net Expansion Income after Taxes is equal to cash inflow.
- End-of-year Cash on Hand—This value is the Beginning-of-year Cash on Hand plus the Net Expansion Income after Taxes.

Internal Rate of Return Calculation Section

This section, as shown in Figure 6-6, helps you use Excel's built-in IRR function.

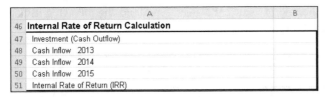

FIGURE 6-6 Internal Rate of Return Calculation section

- Investment (Cash Outflow)—This value is the investment amount from cell B30 in the Calculations section multiplied by –1. The investment value must be a *negative* number to represent it as a cash outflow. (Think of it as money out of your pocket.)
- Cash Inflow 2013—This value is the Net Expansion Income After Taxes for 2013.
- Cash Inflow 2014—This value is the Net Expansion Income After Taxes for 2014.
- Cash Inflow 2015—This value is the Net Expansion Income After Taxes for 2015.
- Internal Rate of Return (IRR)—This value is the annual rate of return that the ranch expansion will generate for Buffalo Ben's. To calculate the IRR, click cell B51, which is where you want to store the IRR result. Then, click the f_x symbol next to the cell-editing window below the Ribbon. The Insert Function dialog box appears (see Figure 6-7). Type IRR in the "Search for a function" text box and then click Go.

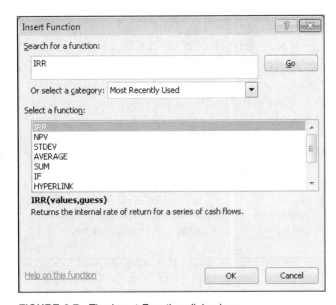

FIGURE 6-7 The Insert Function dialog box

Click OK. The Function Arguments dialog box appears; use it for assistance in building the formula (see Figure 6-8). In the Values text box, either enter the cell references that contain all your cash outflows and inflows (B47:B50), or click and drag your mouse pointer to select cells B47 through B50. Notice that Excel writes the formula for you in cell B51: =IRR(B47:B50). You do not have to enter a value in the Guess text box. When you click OK, Excel calculates the IRR and places the result in cell B51.

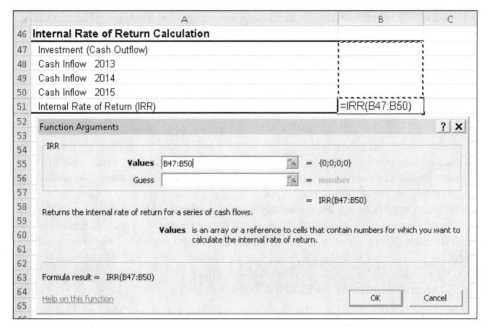

FIGURE 6-8   The Function Arguments dialog box for the IRR function

After you complete the formulas, try testing your spreadsheet with various combinations of the three inputs. There are eight possible combinations, as listed in the next section. If you receive any error messages or see strange values in the cells, go back and check your formulas.

This DSS spreadsheet contains some values that represent millions of dollars. Accountants often simplify their spreadsheets by listing outputs in multiples of thousands or millions of dollars. It is not difficult—you simply divide the cells by one thousand or one million depending on the scale. However, for the purposes of this case, you should keep the large numbers in the spreadsheet. If you see cell results listed as a group of "#" signs (see Figure 6-9), the cell is not wide enough to display the number—widen the column until the number is displayed.

	A	B	C
	Income and Cash Flow Statements	**2012**	**2013**
37			
38	Beginning-of-year Cash on Hand (in 2013 deduct Investment for 2012)	**NA**	########
39	Ranch Expansion Sales Revenues	**NA**	########
40	less: Cost of Goods Sold	**NA**	########
41	Expansion Profit before Income Tax	**NA**	$900,000
42	less: Income Tax Expense	**NA**	$225,000
43	Net Expansion Income after Taxes (Cash Inflow)	**NA**	$675,000
44	End-of-year Cash on Hand	$2,500,000	########

FIGURE 6-9   Column C is not wide enough to display the numbers

ASSIGNMENT 2: USING THE SPREADSHEET FOR DECISION SUPPORT

Next, you will use the spreadsheet to gather data needed to determine the best investment decision and to document your recommendations in a report to Ben and his staff.

As stated previously, this DSS model has eight possible financial outcomes:

1. Buffalo expansion (B)
   a. Normal rainfall and stable economy (N/S/B)
   b. Normal rainfall and recession (N/R/B)
   c. Drought and stable economy (D/S/B)
   d. Drought and recession (D/R/B)
2. Ostrich expansion (O)
   a. Normal rainfall and stable economy (N/S/O)
   b. Normal rainfall and recession (N/R/O)
   c. Drought and stable economy (D/S/O)
   d. Drought and recession (D/R/O)

You are primarily interested in the ranch's financial position based on each of these possible outcomes. The Summary of Key Results section shows the net income after taxes for each of the first three years of expansion, the end-of-year cash on hand at the end of each of the three years, and the internal rate of return. Ben wants to make sure that the selected expansion does not exhaust the cash on hand at the end of the third year (2015). In addition, he would prefer the expansion selection that provides a positive rate of return on investment, regardless of whether there is a drought or a recession. (Recall the two drought years in which the ranch posted a net loss of income.)

Because there are only eight (2^3) possible combinations of inputs among economic outlooks, rainfall and drought scenarios, and expansion selections, you might want to run the spreadsheet model eight times, changing the inputs according to the preceding list. You should do this to ensure that no single year from 2013 through 2015 has negative income or negative end-of-year cash on hand (in other words, to make sure the ranch does not run out of money in any scenario).

You could then transcribe those results to a summary sheet if the management team were interested in the financial results for each year. However, you know that the management team is primarily interested in the financial data from the end of the third year of the model (2015). You can summarize the data easily using Excel Scenario Manager.

Assignment 2A: Using Scenario Manager to Gather Data

For each of the eight situations listed earlier, you want to know the end-of-year cash on hand for the third year of the project (2015), as well as the internal rate of return generated by the cash inflows over the three years.

You will run "what-if" scenarios with the eight sets of input values using Scenario Manager. If necessary, review Tutorial C for tips on using Scenario Manager. In this case, the input values are stored together in one vertical group of cells in the Inputs section (C20 through C22), as are the three output cells in the Summary of Key Results section (E25, E26, and E27), so selecting the cells will be easy. Run Scenario Manager to gather your data in a report called the Scenario Summary. Format the Scenario Summary to make it presentable and then print it for your instructor. Make sure to save your completed Excel workbook before closing it.

Assignment 2B: Documenting Your Recommendations in a Report

Use Microsoft Word to write a brief report to Ben. State the results of your analysis and recommend how to expand the ranch (buffalo or ostrich). Your report must meet the following requirements:

- The first paragraph must summarize the expansion choices and state the purpose of the analysis.
- Next, summarize the results of your analysis and state your recommendation.
- Support your recommendation with a table that outlines the Scenario Summary results. You can use the format shown in Figure 6-10; this table was created in Microsoft Word.
- If the Scenario Summary is well formatted, you might choose to embed an Excel object of it in the body of your report. Tutorial C describes how to copy and paste Excel objects.
- Your instructor might request a graph of the internal rates of return for the eight possible combinations of inputs.

Expansion Selection	Rainfall	Economic Outlook	2015 Net Income ($ millions)	2015 End-of-year Cash on Hand ($ millions)	Internal Rate of Return
Buffalo	Normal	Stable			
		Recession			
	Drought	Stable			
		Recession			
Ostrich	Normal	Stable			
		Recession			
	Drought	Stable			
		Recession			

FIGURE 6-10   Recommended table format for your report

ASSIGNMENT 3: GIVING AN ORAL PRESENTATION WITH SLIDES

Your instructor may ask you to summarize your analysis in an oral presentation. If so, assume that the management team at Buffalo Ben's wants you to explain your analysis and recommendations in 10 minutes or less. A well-designed PowerPoint presentation, with or without handouts, is considered appropriate in a business setting. Tutorial F provides excellent tips for preparing and delivering a presentation.

DELIVERABLES

Your completed case should include the following deliverables for the instructor:

1. A printed copy of your report to management
2. Printouts of your spreadsheets
3. Electronic copies of all your work, including your report, PowerPoint presentation, and Excel DSS model. Ask your instructor which items you should submit for grading.

THE FRUIT PIE INVESTMENT DECISION

Decision Support Using Excel

PREVIEW

You make a great fruit pie, and are considering going into the pie-making business. This business would require investing in a plant and machinery and hiring workers. In this case, you will use Excel to see if the prospective business can be profitable enough to warrant the investment.

PREPARATION

- Review spreadsheet concepts discussed in class and in your textbook.
- Complete any exercises that your instructor assigns.
- Complete any part of Tutorial C that your instructor assigns. You may need to review the use of IF statements and the section called "Cash Flow Calculations: Borrowing and Repayments."
- Review file-saving procedures for Windows programs.
- Refer to Tutorials E and F as necessary.

BACKGROUND

Your mother made delicious apple and cherry pies from scratch, and she left you the recipes. You are proud to say that your pies taste just as good as your mom's! Your family members and friends rave about them. Everyone asks for your recipe, which of course you do not divulge.

As a hobby and part-time business, you make a dozen pies every Friday and sell them to a local bakery. The bakery's customers have come to expect this, and all your pies are usually sold in a couple of hours. This success has made you and your husband think that you could try the fruit pie business on a larger scale.

Your husband is recently retired at the still-young age of 50. He had varied business experiences, and he has figured out what resources are needed to produce and sell pies on a larger scale. He has found a likely production site in town, some refurbished pie-making machinery, and delivery trucks. The plant, equipment, and beginning working capital could be financed by some of your own funds and a $50,000 loan that your local banker says would be available.

In addition, you have taken samples of your work to local bakeries and grocery stores, and many have offered to sell your pies. You think you could sell whatever number of pies you make each day. The question is whether you think a profit can be made.

Your pies feature a delicious crust made from scratch. Of course, the crust has two parts: the top and the slightly larger bottom. Your filling uses fresh, tart green apples or red cherries. The other ingredients, which include flour, nutmeg, shortening, and water, are fresh and of high quality. You bake a pie for 40 minutes at 425 degrees.

A production oven can bake 10 pies at a time. Production would occur in 45-minute blocks: 2 or 3 minutes to load pies into the oven, 40 minutes for baking, and 2 or 3 minutes to put the baked pies into paper boxes and shelve them. A workday would last 12 hours, which is sixteen 45-minute production blocks. Thus, a single worker could produce 160 pies in a day (16 * 10).

You have five ovens, which is the most your premises could handle. Thus, you could produce up to 800 pies a day using five production workers at a time.

You and your husband would prepare the crust and filling for the day's first batch of pies. Thereafter, while one batch bakes, workers would create the crust and filling for the next batch, and put the pies together for baking. At the end of the day, while the last batch of pies bakes, workers would thoroughly clean the premises.

Your production facility would be open 6 days a week for 50 weeks a year, which would be 300 production days per year. You would not produce pies on Sundays.

You think that a delivery truck can be rigged to hold 100 boxed pies. The truck driver would be required to load the truck and then drive to the bakery or grocery store. The driver would unload the truck and take the pies in. In some cases, the driver would shelve the pies; otherwise, store or bakery workers would perform this task.

On average, you think a truck driver could cover 150 miles per day; most of the mileage would be city driving. Your trucks would get 15 miles per gallon. You would lease the needed trucks for $50 per day each.

Production workers would be paid $12 an hour for a 12-hour day. Truck drivers would be paid $16 per hour for a 12-hour day. Overtime rates would not apply for workdays in which employees worked more than 8 hours.

You and your husband would take care of the office work. You expect fixed administrative expenses of $20,000 per year to pay for renting the premises, heat, power, and so on.

Although you expect demand for your pies to be stimulated by word of mouth, you know that you will need an advertising campaign. Your research shows that such campaigns vary in cost and effectiveness. You expect advertising to affect demand and the price you can charge stores and bakeries. Two possible campaigns have been identified. The more expensive campaign should stimulate sales and selling prices more than the other campaign. Each campaign will be more active—and thus more expensive—in the first year.

As a result of severe weather and droughts around the world, the cost of food has increased in recent years. Thus, you are concerned about the rising cost of ingredients. These cost increases probably could not be fully recouped by increasing your selling prices. In the same way, you are concerned about the rising cost of gasoline, which has been about $4 a gallon recently.

You would open for business if you think you could make a profit by 2015, which would be the third year of operations. You want to use Excel to estimate profits for 2013–2015. Your Excel model needs to account for possible sales levels and for changes in ingredient and gasoline costs. Your model will let you develop "what-if" scenarios with the inputs, see the financial results, and then help you decide if you should go into the pie-making business.

ASSIGNMENT 1: CREATING A SPREADSHEET FOR DECISION SUPPORT

In this assignment, you will produce a spreadsheet that models the business decision. Then, in Assignment 2, you will write a memorandum that documents your analysis and conclusions. In Assignment 3, you will prepare and give an oral presentation of your analysis and conclusions to your banker.

First, you will create the spreadsheet model of the financial situation. The model covers the three years from 2013 to 2015. This section helps you set up each of the following spreadsheet components before entering cell formulas:

- Constants
- Inputs
- Summary of Key Results
- Calculations
- Income and Cash Flow Statements
- Debt Owed

A discussion of each section follows. The spreadsheet skeleton is available for you to use. To access this skeleton, go to your data files, select Case 7, and then select **Pies.xlsx**.

Constants Section

Your spreadsheet should include the constants shown in Figure 7-1. An explanation of the line items follows the figure.

	A	B	C	D	E
1	**The Fruit Pie Investment Decision**				
2					
3	**Constants**	**2012**	**2013**	**2014**	**2015**
4	Tax Rate	NA	20%	20%	20%
5	Minimum cash needed to start year	NA	$ 10,000	$ 10,000	$ 10,000
6	Number of production days	NA	300	300	300
7	Maximum number of pies produced per day	NA	800	800	800
8	Number of pies a truck can carry	NA	100	100	100
9	Cost to lease a truck per day	NA	$ 50	$ 50	$ 50
10	Number of truck miles per day	NA	150	150	150
11	Truck miles per gallon	NA	15	15	15
12	Production worker pay per hour	NA	$ 12	$ 12	$ 12
13	Truck driver pay per hour	NA	$ 16	$ 16	$ 16
14	Administrative costs in year	NA	$ 20,000	$ 20,000	$ 20,000
15	High cost advertising campaign	NA	$ 20,000	$ 12,000	$ 6,000
16	Low cost advertising campaign	NA	$ 10,000	$ 7,000	$ 4,000
17	Interest rate on debt owed	NA	4%	5%	6%

FIGURE 7-1 Constants section

- Tax Rate—The tax rate is applied to income before taxes. The rate is expected to be the same each year.
- Minimum cash needed to start year—You want to have at least $10,000 in cash at the beginning of each year. Your banker will lend you the needed amount at the end of a year to begin the new year with $10,000.
- Number of production days—Your production facility will operate 300 days a year.
- Maximum number of pies produced per day—You can make 800 pies per day at most.
- Number of pies a truck can carry—A truck can be set up to comfortably store 100 pies on racks.
- Cost to lease a truck per day—A truck can be leased for $50 a day.
- Number of truck miles per day—On average, a driver can cover 150 miles per day in and around the city.
- Truck miles per gallon—A truck is expected to average 15 miles per gallon.
- Production worker pay per hour—The rate is $12 an hour, with no overtime.
- Truck driver pay per hour—The rate is $16 an hour, with no overtime.
- Administrative costs in year—Fixed costs are expected to be $20,000 per year for rent, maintenance, insurance, electricity, and so on. These costs do not fluctuate with changes in production.
- High cost advertising campaign—The more expensive campaign will cost $20,000 in the first year, with costs declining as the years go on.
- Low cost advertising campaign—The less expensive campaign will cost $10,000 in the first year, with costs declining as the years go on.
- Interest rate on debt owed—Your banker will charge interest for any debt owed in the year. The rate can vary; the banker says that interest rates are expected to rise as the economy continues to recover.

Inputs Section

Your spreadsheet should include the following inputs for the years 2013, 2014, and 2015, as shown in Figure 7-2.

	A	B	C	D	E
19	**Inputs**	**2012**	**2013**	**2014**	**2015**
20	Advertising Level (1=High, 2=Low)		NA	NA	NA
21	Food inflation in year (.xx)	NA			
22	Cost of gallon of gas in year ($x.xx)	NA			

FIGURE 7-2 Inputs section

- Advertising Level (1=High, 2=Low)—Enter the number that corresponds to the selected campaign.
- Food inflation in year (.xx)—In each year, enter the rate of inflation expected. For example, if 5% inflation is expected in 2013, enter .05 in cell C21. Format these cells as a percentage.
- Cost of gallon of gas in year ($x.xx)—In each year, enter the expected cost of a gallon of gasoline. For example, if you expect a gallon to cost $4.50 in 2013, enter 4.50 in cell C22. Format these cells as currency.

Summary of Key Results Section

Your spreadsheet should include the results shown in Figure 7-3.

	A	B	C	D	E
24	**Summary of Key Results**	**2012**	**2013**	**2014**	**2015**
25	Net income after taxes	NA			
26	End-of-year cash on hand	NA			
27	End-of-year debt owed	NA			

FIGURE 7-3 Summary of Key Results section

For each year, your spreadsheet should show net income after taxes, cash on hand at the end of the year, and debt owed to the bank at the end of the year. The net income, cash, and debt cells should be formatted as currency with zero decimal places. These values are computed elsewhere in the spreadsheet and should be echoed here.

Calculations Section

You should calculate intermediate results that will be used in the income and cash flow statements that follow. Calculations, as shown in Figure 7-4, may be based on expected year-end 2012 values. When called for, use absolute referencing properly. Values must be computed by cell formula; hard-code numbers in formulas only when you are told to do so. Cell formulas should not reference a cell with a value of "NA."

An explanation of each item in this section follows the figure.

	A	B	C	D	E
29	**Calculations**	**2012**	**2013**	**2014**	**2015**
30	Per day demand for pies	NA			
31	Pies produced in year	NA			
32	Number of production workers needed	NA			
33	Number of delivery trucks needed	NA			
34	Truck miles per day	NA			
35	Gallons of gas used per day	NA			
36	Production worker pay	NA			
37	Truck driver pay	NA			
38	Cost of gasoline	NA			
39	Selling price of a pie	NA			
40	Variable cost of a pie	$ 6.00			
41	Advertising cost	NA			
42	Truck leasing cost	NA			

FIGURE 7-4 Calculations section

- Per day demand for pies—This value is the number of pies that stores and bakeries want to buy each day. The amount is a function of the advertising level. In 2013, if the more expensive campaign is used, 480 pies per day will be needed; otherwise, 320 pies per day will be needed. In 2014, if the more expensive campaign is used, the demand will be twice the 2013 demand; otherwise, the demand will be 1.5 times the 2013 demand. In 2015, if the more expensive campaign is used, the demand will be 1.5 times the 2014 demand; otherwise, the demand will be 1.25 times the 2014 demand. However, the demand cannot exceed the constant 800; you can deal with this constraint by using the MIN() function or an IF() statement.
- Pies produced in year—This amount is a function of the number of pies needed per day and the number of production days, which is a value from the Constants section.

- Number of production workers needed—This amount is a function of the daily demand and the output per worker, 160, a value you can hard-code in your formula. Use the ROUND() function with zero decimal places to avoid specifying a part of a worker.
- Number of delivery trucks needed—This amount is a function of the daily demand and truck capacity, a constant value. Use the ROUND() function with zero decimal places to avoid specifying a part of a truck.
- Truck miles per day—This amount is a function of the number of trucks and truck mileage per day from the Constants section.
- Gallons of gas used per day—This amount is a function of truck mileage per day and miles per gallon from the Constants section. Use the ROUND() function with zero decimal places to avoid specifying a part of a gallon.
- Production worker pay—This amount is the production worker pay for all workers during the year. The value is a function of the number of workers, the pay per hour, the number of hours per day (12, a constant you can hard-code), and the number of business days.
- Truck driver pay—This amount is the truck driver pay for all drivers during the year. The value is a function of the number of drivers, the pay per hour, the number of hours per day (12, a constant you can hard-code), and the number of business days.
- Cost of gasoline—This amount is a function of the number of gallons used per day, the cost of a gallon of gasoline during the year (a value in the Inputs section), and the number of business days.
- Selling price of a pie—The price is $11 each year if the more expensive advertising campaign is used; otherwise, the price is $10.
- Variable cost of a pie—This amount is the cost of ingredients, supplies, power, and other items used to make a pie. The cost is a function of the food price increase during the year and the prior year's variable cost. Thus, the 2013 cost is based on the 2012 cost and the 2013 food cost increase. The 2014 cost is based on the 2013 cost and the 2014 food cost increase, and so on.
- Advertising cost—This amount is a function of the campaign value from the Inputs section. Campaign costs are shown in the Constants section.
- Truck leasing cost—This amount is a function of the number of trucks, the leasing cost per day, and the number of business days.

Income and Cash Flow Statements

The forecast for net income and cash flow starts with the cash on hand at the beginning of the year. This value is followed by the income statement and the calculation of cash on hand at year's end. For readability, format cells in this section as currency with zero decimal places. Values must be computed by cell formula; hard-code numbers in formulas only if you are told to do so. Cell formulas should not reference a cell with a value of "NA." Your spreadsheets should look like those in Figures 7-5 and 7-6. A discussion of each item in the section follows each figure.

	A	B	C	D	E
44	**Income and Cash Flow Statements**	**2012**	**2013**	**2014**	**2015**
45	Beginning-of-year cash on hand	NA			
46					
47	Revenue	NA			
48	Costs:				
49	Cost of pies produced	NA			
50	Production worker pay	NA			
51	Truck driver pay	NA			
52	Truck leasing	NA			
53	Advertising	NA			
54	Gasoline	NA			
55	Administrative	NA			
56	Total costs	NA			
57	Income before interest and taxes	NA			
58	Interest expense	NA			
59	Income before taxes	NA			
60	Income tax expense	NA			
61	Net income after taxes	NA			

FIGURE 7-5 Income and Cash Flow Statements section

- Beginning-of-year cash on hand—This value is the cash on hand at the end of the prior year.
- Revenue—This amount is a function of the number of pies needed and the selling price. Both of these values are from the Calculations section.
- Cost of pies produced—This value is a function of the number of pies produced during the year and the variable cost of a pie. Both of these values are taken from the Calculations section.
- Production worker pay—This amount from the Calculations section can be echoed here.
- Truck driver pay—This amount from the Calculations section can be echoed here.
- Truck leasing—This amount from the Calculations section can be echoed here.
- Advertising—This amount from the Calculations section can be echoed here.
- Gasoline—This amount from the Calculations section can be echoed here.
- Administrative—This amount from the Constants section can be echoed here.
- Total costs—This amount is the total cost of pie production, production worker pay, truck driver pay, truck leasing, advertising, gasoline, and administrative costs.
- Income before interest and taxes—This amount is the difference between revenue and total costs.
- Interest expense—This amount is the product of the debt owed to the bank at the beginning of the year and the interest rate for the year (a value from the Constants section).
- Income before taxes—This amount is the difference between income before interest and taxes and interest expense.
- Income tax expense—This amount is zero if income before taxes is zero or less than zero (a loss). Otherwise, income tax expense is the product of the year's tax rate (a value from the Constants section) and income before taxes.
- Net income after taxes—This amount is the difference between income before taxes and income tax expense.

Line items for the year-end cash calculation are discussed next. In Figure 7-6, Column B represents 2012, Column C is for 2013, and so on. Year 2012 values are NA except for End-of-year cash on hand, which is $10,000.

	A	B	C	D	E
63	Net cash position (NCP)	NA			
64	Borrowing from bank	NA			
65	Repayment to bank	NA			
66	End-of-year cash on hand	$ 10,000			

FIGURE 7-6 End-of-year cash on hand section

- Net cash position (NCP)—The NCP at the end of a year equals the cash at the beginning of the year plus the year's net income after taxes.
- Borrowing from bank—Assume that your bank will lend enough money at the end of the year to reach the minimum cash needed to start the next year. If the NCP is less than this minimum, the company must borrow enough to start the next year with the minimum. Borrowing increases the cash on hand, of course.
- Repayment to bank—If the NCP is more than the minimum cash needed and some debt is owed at the beginning of the year, you must pay off as much debt as possible (without taking the cash below the minimum amount required to start the next year). Repayments reduce cash on hand, of course.
- End-of-year cash on hand—This amount is the NCP plus any borrowing and minus any repayments.

Debt Owed Section

This section shows a calculation of debt owed to the bank at year's end, as shown in Figure 7-7. Year 2012 values are NA except for End-of-year debt owed, which is $50,000. Values must be computed by cell formula; hard-code numbers in formulas only when you are told to do so. Cell formulas should not reference a cell with a value of "NA." An explanation of each item follows the figure.

	A	B	C	D	E
68	**Debt Owed**	**2012**	**2013**	**2014**	**2015**
69	Beginning-of-year debt owed	NA			
70	Borrowing from bank	NA			
71	Repayment to bank	NA			
72	End-of-year debt owed	$ 50,000			

FIGURE 7-7 Debt Owed section

- Beginning-of-year debt owed—Debt owed at the beginning of a year equals the debt owed at the end of the prior year.
- Borrowing from bank—This amount has been calculated elsewhere and can be echoed to this section. Borrowing increases the amount of debt owed.
- Repayment to bank—This amount has been calculated elsewhere and can be echoed to this section. Repayments reduce the amount of debt owed.
- End-of-year debt owed—In all years, this is the amount owed at the beginning of a year, plus borrowing during the year, and minus repayments during the year.

ASSIGNMENT 2: USING THE SPREADSHEET FOR DECISION SUPPORT

Complete the case by (1) using the spreadsheet to gather data needed to determine if you can make money in the pie business, and (2) documenting your findings in a memorandum.

You want to model three economic combinations of food price inflation and gasoline price. In the Favorable condition, food inflation is only 2% each year and a gallon of gasoline costs $3.50 each year. In the SoSo condition, food inflation is 5% each year and a gallon of gasoline costs $4.50 each year. In the Unfavorable condition, food inflation is 10% each year and a gallon of gasoline costs $5.00 each year.

Within each of these three conditions, there are two possible advertising levels: high and low. The resulting six scenarios are summarized as follows:

1. Favorable-High—This scenario combines the Favorable economic condition and the High advertising campaign.
2. Favorable-Low—This scenario combines the Favorable economic condition and the Low advertising campaign.
3. SoSo-High—This scenario combines the SoSo economic condition and the High advertising campaign.
4. SoSo-Low—This scenario combines the SoSo economic condition and the Low advertising campaign.
5. Unfavorable-High—This scenario combines the Unfavorable economic condition and the High advertising campaign.
6. Unfavorable-Low—This scenario combines the Unfavorable economic condition and the Low advertising campaign.

You want answers to the following questions:

- In the Favorable economic condition, will the company have a positive 2015 net income after taxes, no matter what advertising level is chosen?
- In the SoSo economic condition, will the company have a positive 2015 net income after taxes, no matter what advertising level is chosen?
- In the Unfavorable economic condition, will the company have a positive 2015 net income after taxes with either advertising level?
- If you can assume that the Unfavorable economic condition is very unlikely to occur, what advertising strategy (or strategies) should be used?

You will use your spreadsheet model to gather data needed to answer these questions.

Assignment 2A: Using the Spreadsheet to Gather Data

You have built the spreadsheet to model the business situation. For each of the six scenarios, you want to know the 2015 net income after taxes, the 2015 end-of-year cash on hand, and the 2015 end-of-year debt owed.

You will run "what-if" scenarios with the six sets of input values using Scenario Manager. (See Tutorial C for details on using Scenario Manager.) Set up the six scenarios. Your instructor may ask you to use conditional formatting to make sure that your input values are proper. (Note that in Scenario Manager you can enter noncontiguous cell ranges, such as C19, D19, and C20:F20.)

The relevant output cells are the three 2015 cells in the Summary of Key Results section. Run Scenario Manager to gather the data in a report sheet. When you finish, print the spreadsheet with the input for any of the scenarios, print the Scenario Manager summary sheet, and then save the spreadsheet file for the last time.

Assignment 2B: Documenting Your Recommendations in a Memorandum

Use Microsoft Word to write a brief memorandum documenting your analysis and conclusions. You can address the memo to your banker. Observe the following requirements:

- Set up your memo as described in Tutorial E.
- In the first paragraph, briefly state the business situation and the purpose of your analysis.
- Next, provide the answers to your four questions.
- State your recommendation: Should you enter the pie-making business?
- Support your statements graphically, as your instructor requires. Your instructor may ask you to return to Excel and copy the Scenario Manager summary results into the memo. (See Tutorial C for details on this procedure.) Your instructor might also ask you to make a summary table in Word based on the Scenario Manager summary results. (This procedure is described in Tutorial E.)

Your table should resemble the format shown in Figure 7-8.

Scenario	2015 Net Income	2015 Cash on Hand	2015 Debt Owed
Favorable-High			
Favorable-Low			
SoSo-High			
SoSo-Low			
Unfavorable-High			
Unfavorable-Low			

FIGURE 7-8   Format of table to insert in memo

ASSIGNMENT 3: GIVING AN ORAL PRESENTATION

Your instructor may ask you to explain your analysis and recommendations in an oral presentation. If so, assume that your banker wants the presentation to last 10 minutes or less. Use visual aids or handouts that you think are appropriate. See Tutorial F for tips on preparing and giving an oral presentation.

DELIVERABLES

Your completed case should include the following deliverables for the instructor:

1. A printed copy of your report to the banker
2. Printouts of your spreadsheets
3. Electronic copies of all your work, including your report, PowerPoint presentation, and Excel DSS model. Ask your instructor which items you should submit for grading.

PART 3

DECISION SUPPORT CASES USING MICROSOFT EXCEL SOLVER

BUILDING A DECISION SUPPORT SYSTEM USING MICROSOFT EXCEL SOLVER

In Tutorial C, you learned that Decision Support Systems (DSS) are programs used to help managers solve complex business problems. Cases 6 and 7 were DSS models that used Microsoft Excel Scenario Manager to calculate and display financial outcomes given certain inputs, such as economic outlooks and mortgage interest rates. You used the outputs from Scenario Manager to see how different combinations of inputs affected cash flows and income so that you could make the best decision for expanding your business or selecting a technology to develop and market.

Many business situations require models in which the inputs are not limited to two or three choices, but include large ranges of numbers in more than three variables. For such business problems, managers want to know the best or optimal solution to the model. An optimal solution can either maximize an objective variable, such as income or revenues, or minimize the objective variable, such as operating costs. The formula or equation that represents the target income or operating cost is called an objective function. Optimizing the objective function requires the use of constraints (also called constraint equations), which are rules or conditions you must observe when solving the problem. The field of applied mathematics that addresses problem solving with objective functions and constraint equations is called linear programming. Before the advent of digital computers, linear programming required the knowledge of complex mathematical techniques. Fortunately, Excel has a tool called Solver that can compute the answers to optimization problems.

This tutorial has five sections:

1. **Adding Solver to the Ribbon**—Solver is not installed by default with Excel 2010; you must add it to the application. You may need to use Excel Options to add Solver to the Ribbon.
2. **Using Solver**—This section explains how to use Solver. You will start by determining the best mix of vehicles for shipping exercise equipment to stores throughout the country.
3. **Extending the example**—This section tests your knowledge of Solver as you modify the transportation mix to accommodate changes: additional stores to supply and redesign of the product to reduce shipping volume.
4. **Using Solver on a new problem**—In this section, you will use Solver on a new problem: maximizing the profits for a mix of products.
5. **Troubleshooting Solver**—Because Solver is a complex tool, you will sometimes have problems using it. This section explains how to recognize and overcome such problems.

NOTE

If you need a refresher, Tutorial C offers guidance on basic Excel concepts such as formatting cells and using the =IF( ) and AND( ) functions.

ADDING SOLVER TO THE RIBBON

Before you can use Solver, you must determine whether it is installed in Excel. Start Excel and then click the Data tab on the Ribbon. If you see a group on the right side named Analysis that contains Solver, you do not need to install Solver (see Figure D-1).

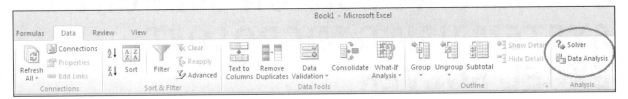

FIGURE D-1 Analysis group with Solver installed

If the Analysis group or Solver is not shown on the Data tab of the Ribbon, do the following:

1. Click the File tab.
2. Click Options (see Figure D-2).
3. Click Add-Ins (see Figure D-3) to display the available add-ins in the right pane.
4. Click Go at the bottom of the right pane. The window shown in Figure D-4 appears.
5. Click the Solver Add-in box as well as the Analysis ToolPak and Analysis ToolPak-VBA boxes. (You will need the latter options in a subsequent case, so install them now with Solver.)
6. Click OK to close the window and return to the Ribbon. If you click the Data tab again, you should see the Analysis group with Data Analysis and Solver on the right.

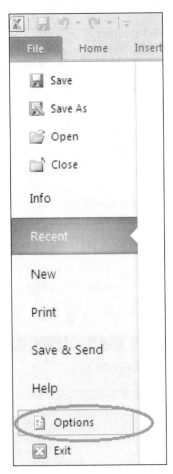

FIGURE D-2 Excel Options selection

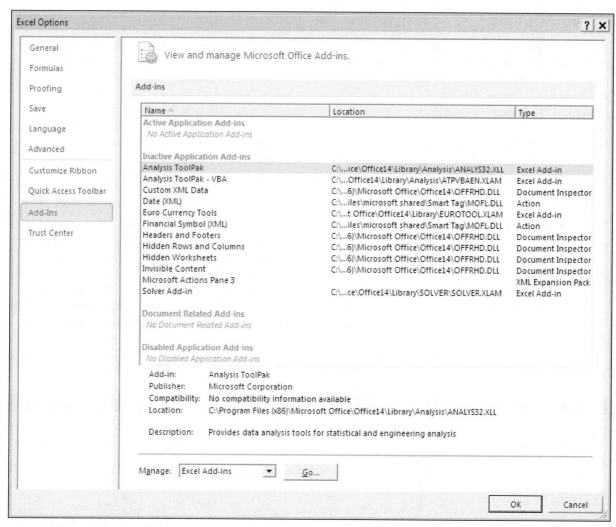

FIGURE D-3 Add-Ins pane

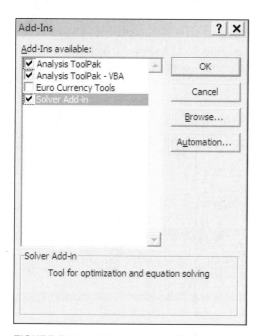

FIGURE D-4 Add-Ins window with Solver, Analysis ToolPak, and Analysis ToolPak VBA selected

USING SOLVER

A fictional company called CV Fitness builds exercise machines in its plant in Memphis, Tennessee and ships them to its stores across the country. The company has a small fleet of trucks and tractor-trailers to ship its products from the factory to its stores. It costs less money per cubic foot of capacity to ship products with tractor-trailers than with trucks, but the company has a limited number of both types of vehicles and must ship a specified amount of each type of product to each destination. You have been asked to determine the optimal mix of trucks and tractor-trailers to send merchandise to each store. The optimal mix will have the lowest total shipping cost while ensuring that the required quantity of products is shipped to each store.

To use Solver, you must set up a model of the problem, including the factors that can vary (the mix of trucks and tractor-trailers) and the constraints on how much they can vary (the number of each vehicle available). Your goal is to minimize the shipping cost.

Setting Up a Spreadsheet Skeleton

CV Fitness makes three fitness machines: exercise bikes (EB), elliptical cross-trainers (CT), and treadmills (TM). When packaged for shipment, their shipping volumes are 12, 15, and 22 cubic feet, respectively. The finished machines are shipped via ground transportation to five stores in Philadelphia, Atlanta, Miami, Chicago, and Los Angeles. Your vehicle fleet consists of 12 trucks and six tractor-trailers. Each truck has a capacity of 1500 cubic feet, and each tractor-trailer has a capacity of 2350 cubic feet. The spreadsheet includes the road distances from your plant in Memphis to each store, along with each store's demand for the three fitness machines.

What is the best mix of trucks and tractor-trailers to send to each destination? You will learn how to use Solver to determine the answer. The spreadsheet components are discussed in the following sections.

AT THE KEYBOARD

Start by saving your blank spreadsheet. Use a descriptive filename so you can find it easily later—**CV Fitness Trucking Problem.xlsx** should work well. Then enter the skeleton and formulas as directed in the following sections.

Spreadsheet Title

Resize Column A, as illustrated in Figure D-5, to give your spreadsheet a small border on the left side. Enter the spreadsheet title in cell B1. Merge and center cells B1 through F1 using the Merge and Center button in the Alignment group of the Home tab.

Constants Section

Your spreadsheet should have a section for values that will not change. Figure D-5 shows a skeleton of the Constants section and the values you should enter. A discussion of the line items follows the figure.

	A	B	C	D	E	F
1		CV Fitness, Inc. Truck Load Management Problem				
2						
3		Constants Section:				
4			Volume Cu. Ft.	Operating Cost per mi.	Operating Cost per mi-cu. Ft.	Available Fleet
5		Truck	1500	$1.00	$0.000667	12
6		Tractor Trailer	2350	$1.30	$0.000553	6
7						
8		Exercise Bike (EB)	12			
9		Elliptical Crosstrainer (CT)	15			
10		Treadmill (TM)	22			

FIGURE D-5   Spreadsheet title and Constants section

- In column C, enter the Volume Cu. Ft., which is the cubic-foot capacity of the vehicles as well as the shipping volume for each item of exercise equipment.
- In column D, enter the Operating Cost per mi., which is the cost per mile driven for each type of vehicle.
- In column E, enter the Operating Cost per mi.-cu.ft. This value is actually a formula: the operating cost per mile divided by the vehicle volume in cubic feet. Normally you do not put formulas in the Constants section, but in this case it lets you see the relative cost efficiencies of each vehicle. Assuming that both types of vehicles can be filled to capacity, the tractor-trailer is the preferred vehicle for shipping cost efficiency.
- In column F, enter the values for the Available Fleet, which is the number of each type of vehicle your company owns or leases.

You can update the Constants section as the company adds more products to its offerings or adds vehicles to its fleet.

NOTE

The column headings in the Constants section contain two or three lines to keep the columns from becoming too wide. To create a new line in a cell, hold down the Alt key and press Enter.

Now is a good time to save your workbook again. Keep the name you assigned earlier.

Calculations and Results Section

The structure and format of your Calculations and Results section will vary greatly depending on the nature of the problem you need to solve. In some Solver models, you might need to maximize income, which means you might also have an Income Statement section. In other Solver models, you may want to have a separate Changing Cells section that contains cells Solver will manipulate to obtain a solution. In this tutorial, you want to minimize shipping costs while meeting the product demand of your stores. You can accomplish this task by building a single unified table that includes the distances to the stores, the product demand for each store, and the shipping alternatives and costs.

A unified Calculations and Results section makes sense in this model for several reasons. First, it simplifies writing and copying the formulas for the needed shipping volumes, the vehicle capacity totals, and the shipping costs to each destination. Second, a well-organized table allows you to easily identify the changing cells, which Solver will manipulate to optimize the solution, as well as the total cost (or optimization cell). Finally, a unified table allows your management team to visualize both the problem and its solution.

When creating a complex table, it is often a good idea to sketch the table's structure first to see how you want to organize the data. Format the table structure, then enter the data you are given for the problem. Write the cells that contain the formulas last, starting with all the formulas in the first row. If you do a good job structuring your table, you will be able to copy the first-row formulas to the other rows.

Build the blank table shown in Figure D-6. A discussion of the rows and columns follows the figure.

NOTE

Leave rows 11 and 12 blank between the Constants section and the Calculations and Results section. You then will have room to add an extra product to your Constants section later.

	Calculations and Results Section:												
	Distance/Demand Table			Store Demand			Vehicle Loading							Cost
	Distance Table (from Memphis Plant)	Miles	EB	CT	TM	Volume Required	Trucks	Volume for Trucks	Tractor-Trailers	Volume for Tractor-Trailers	Total Vehicle Capacity	% of Vehicle Capacity Utilized	Shipping Cost
Philadelphia Store	1010	140	96	86									
Atlanta Store	380	76	81	63									
Miami Store	1000	56	64	52									
Chicago Store	540	115	130	150									
Los Angeles Store	1810	150	135	180									
					Totals:								
Fill Legend:			Changing Cells										Total Cost
			Optimization Cell										

FIGURE D-6   Blank table for Calculations and Results section

- In row 13, enter "Calculations and Results Section:" as the title of the table.
- In row 14, columns B and C, enter "Distance/Demand Table" as a column heading. Merge and center the heading in the two columns.
- In row 14, columns D, E, and F, enter "Store Demand" as a column heading. Merge and center the heading in the three columns.
- In row 14, columns G through M, enter "Vehicle Loading" as a column heading. Merge and center the heading across the columns.
- In row 14, column N, enter "Cost" as a centered column heading.
- In row 15, column B, enter "Distance Table (from Memphis Plant)" as a centered column heading.
- In row 15, column C, enter "Miles" as a centered column heading.
- In row 15, columns D, E, and F, enter "EB," "CT," and "TM," respectively, as equipment headings.
- In row 15, columns G through N, enter "Volume Required," "Trucks," "Volume for Trucks," "Tractor-Trailers," "Volume for Tractor-Trailers," "Total Vehicle Capacity," "% of Vehicle Capacity Utilized," and "Shipping Cost," respectively, as column headings.
- In rows 16 through 20, column B, enter the destination store locations.
- In rows 16 through 20, column C, enter the number of miles to the destination store locations.
- In rows 16 through 20, columns D through F, enter the number of exercise bikes (EB), cross-trainers (CT), and treadmills (TM) to be shipped to each store location.
- Rows 16 through 20, columns G through N, will contain formulas or "seed values" later. Leave them blank for now, but fill cells H16 through H20 and cells J16 through J20 with a light color to indicate that they are the changing cells for Solver. To fill a cell, use the Fill Color button in the Font group.
- In cell F21, enter "Totals:" to label the following cells in the row.
- Cells G21 through N21 will be used for column totals. Fill cell N21 with a slightly darker shade than you used for the changing cells. Cell N21 is your optimization cell.
- In cell B22, enter "Fill Legend:" as a label.
- Fill cell C22 with the fill color you selected for the changing cells.
- In cells D22 and E22, enter "Changing Cells" as the label for the fill color. Merge and center the label in the cells.
- In cell N22, enter "Total Cost" as the label for the value in cell N21.
- Fill cell C23 with the fill color you selected for the optimization cell.
- In cells D23 and E23, enter "Optimization Cell" as the label for the fill color. Merge and center the label in the cells.

Figure D-7 illustrates a magnified section of the Distance/Demand table in case the numbers in Figure D-6 are difficult to read.

	A	B	C	D	E	F
12						
13		Calculations and Results Section:				
14		Distance/Demand Table		Store Demand		
15		Distance Table (from Memphis Plant)	Miles	EB	CT	TM
16		Philadelphia Store	1010	140	96	86
17		Atlanta Store	380	76	81	63
18		Miami Store	1000	56	64	52
19		Chicago Store	540	115	130	150
20		Los Angeles Store	1810	150	135	180
21						Totals:
22		Fill Legend:		Changing Cells		
23				Optimization Cell		

FIGURE D-7   Magnified view of the Distance/Demand table

Use the Borders menu in the Font group to select and place appropriate borders around parts of the Calculations and Results section (see Figure D-8). The All Borders and Outside Borders selections are the most useful borders for your table.

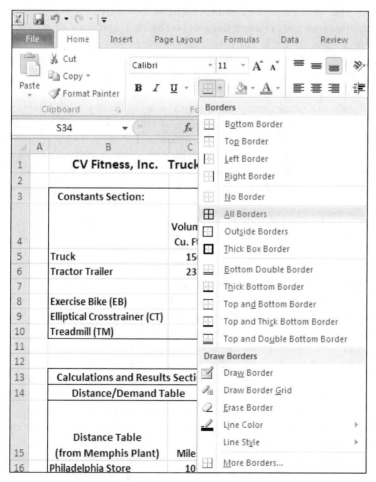

FIGURE D-8   Borders menu

Next, you write the formulas for the volume and cost calculations. Figure D-9 shows a magnified view of the Vehicle Loading and Cost sections. A discussion of the formulas required for the cells follows the figure.

	G	H	I	J	K	L	M	N
13								
14			Vehicle Loading					Cost
15	Volume Required	Trucks	Volume for Trucks	Tractor-Trailers	Volume for Tractor-Trailers	Total Vehicle Capacity	% of Vehicle Capacity Utilized	Shipping Cost
16								
17								
18								
19								
20								
21								
22								Total Cost

FIGURE D-9   Vehicle Loading and Cost sections

For illustration purposes, the cell numbers in the following list refer to values for the Philadelphia store.

- Volume Required—Cell G16 contains the total shipping volume of the three types of equipment shipped to the Philadelphia store. The formula for this cell is =D16*$C$8+E16*$C$9+F16*$C$10. Cells D16, E16, and F16 are the quantities of each item to be shipped, and cells $C$8, $C$9, and $C$10 are the shipping volumes for the exercise bike, cross-trainer, and treadmill, respectively. When taking values from the Constants section to calculate formulas, you almost always should use absolute cell references ($) because you will copy the formulas down the columns.
- Trucks—Cell H16 contains the number of trucks selected to ship the merchandise. Cell H16 is a changing cell, which means Solver will determine the best number of trucks to use and place the number in this cell. For now, you should "seed" the cell with a value of 1.
- Volume for Trucks—Cell I16 contains the number of trucks selected, multiplied by the capacity of a truck. The capacity value is taken from the Constants section. The formula for this cell is =H16*$C$5. Cell H16 is the number of trucks selected, and cell $C$5 is the volume capacity of the truck in cubic feet.
- Tractor-Trailers—Cell J16 contains the number of tractor-trailers selected to ship the merchandise. Cell J16 is a changing cell, which means Solver will determine the best number of tractor-trailers to use and place the number in this cell. For now, you should "seed" the cell with a value of 1.
- Volume for Tractor-Trailers—Cell K16 contains the number of tractor-trailers selected, multiplied by the capacity of a tractor-trailer. The capacity value is taken from the Constants section. The formula for this cell is =J16*$C$6. Cell J16 is the number of tractor-trailers selected, and cell $C$6 is the cubic feet capacity of the tractor-trailer.
- Total Vehicle Capacity—Cell L16 contains the sum of the Volume for Trucks and the Volume for Tractor-Trailers. The formula for this cell is =I16+K16. You need to know the Total Vehicle Capacity to make sure that you have enough capacity to ship the Volume Required. This value will be one of your constraints in Solver.
- % of Vehicle Capacity Utilized—Cell M16 contains the Volume Required divided by the Total Vehicle Capacity. The formula for this cell is =G16/L16; after entering the formula, format it as a percentage using the % button in the Number group. Although this information is not required to minimize shipping costs, it is useful for managers to know how much space was filled in the selected vehicles. Alternatively, you could run Solver to determine the highest space utilization on the vehicles rather than the lowest cost. Note that you cannot use more than 100% of the available space on the vehicles.

- Shipping Cost—Cell N16 contains the following calculation:

Mileage to destination store × Number of trucks selected × Cost per mile for trucks + Mileage to destination store × Number of tractor-trailers selected × Cost per mile for tractor-trailers

The formula for this cell is =H16*C16*$D$5+J16*C16*$D$6. Note that absolute cell references for the cost-per-mile values are taken from the Constants section.

If you entered the formulas correctly in row 16, your table should look like Figure D-10.

	G	H	I	J	K	L	M	N
14				Vehicle Loading				Cost
15	Volume Required	Trucks	Volume for Trucks	Tractor-Trailers	Volume for Tractor-Trailers	Total Vehicle Capacity	% of Vehicle Capacity Utilized	Shipping Cost
16	5012	1	1500	1	2350	3850	130%	$2,323.00
17								
18								
19								
20								
21								
22								Total Cost

FIGURE D-10   Vehicle Loading and Cost sections with formulas entered in the first row

To complete the empty cells in rows 17 through 20, you can copy the formulas from cells G16 through N16 to the rest of the rows. Click and drag to select cells G16 through N16, then right-click and select Copy from the menu (see Figure D-11).

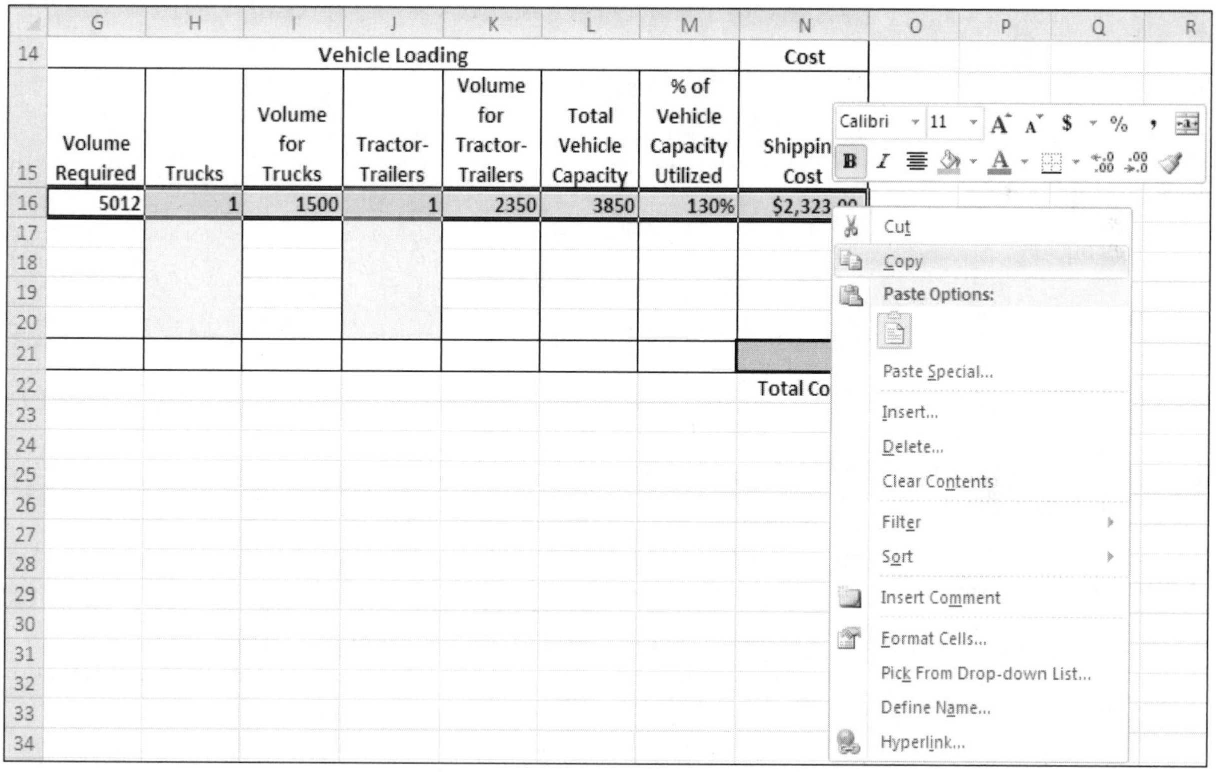

	G	H	I	J	K	L	M	N	O	P	Q	R
14				Vehicle Loading				Cost				
15	Volume Required	Trucks	Volume for Trucks	Tractor-Trailers	Volume for Tractor-Trailers	Total Vehicle Capacity	% of Vehicle Capacity Utilized	Shippin Cost				
16	5012	1	1500	1	2350	3850	130%	$2,323.00				
17									Cut			
18									Copy			
19									Paste Options:			
20												
21									Paste Special...			
22								Total Co	Insert...			
23									Delete...			
24									Clear Contents			
25												
26									Filter	▶		
27									Sort	▶		
28												
29									Insert Comment			
30												
31									Format Cells...			
32									Pick From Drop-down List...			
33									Define Name...			
34									Hyperlink...			

FIGURE D-11   Copying formulas

Next, select cells G17 through N20, which are in the four rows beneath row 16. Either press Enter or click Paste in the Clipboard group. The formulas from row 16 should be copied to the rest of the destination cities (see Figure D-12).

	G	H	I	J	K	L	M	N
14	Vehicle Loading							Cost
15	Volume Required	Trucks	Volume for Trucks	Tractor-Trailers	Volume for Tractor-Trailers	Total Vehicle Capacity	% of Vehicle Capacity Utilized	Shipping Cost
16	5012	1	1500	1	2350	3850	130%	$2,323.00
17	3513	1	1500	1	2350	3850	91%	$874.00
18	2776	1	1500	1	2350	3850	72%	$2,300.00
19	6630	1	1500	1	2350	3850	172%	$1,242.00
20	7785	1	1500	1	2350	3850	202%	$4,163.00
21								
22								Total Cost

FIGURE D-12   Formulas from row 16 successfully copied to rows 17 through 20

You have one row of formulas to complete: the Totals row. You will use the AutoSum function to sum up one column, and then copy the formula to the rest of the columns *except* cell M21. This cell is not actually a total, but an overall capacity utilization rate.

To enter the sum of cells G16 through G20 in cell G21, select cells G16 through G21, then click AutoSum in the Editing group on the Home tab of the Ribbon (see Figure D-13).

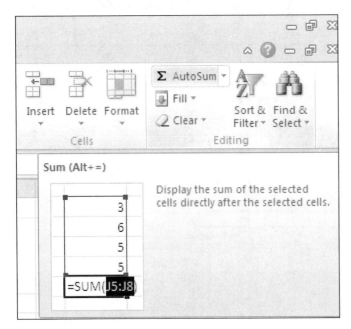

FIGURE D-13   AutoSum button in the Editing group

Cell G21 should now contain the formula =SUM(G16:G20), and the displayed answer should be 25716. Now you can copy cell G21 to cells H21, I21, J21, K21, L21, and N21. When you have completed this section of the table, it should have the values shown in Figure D-14.

	Vehicle Loading						Cost
Volume Required	Trucks	Volume for Trucks	Tractor-Trailers	Volume for Tractor-Trailers	Total Vehicle Capacity	% of Vehicle Capacity Utilized	Shipping Cost
5012	1	1500	1	2350	3850	130%	$2,323.00
3513	1	1500	1	2350	3850	91%	$874.00
2776	1	1500	1	2350	3850	72%	$2,300.00
6630	1	1500	1	2350	3850	172%	$1,242.00
7785	1	1500	1	2350	3850	202%	$4,163.00
25716	5	7500	5	11750	19250		$10,902.00
							Total Cost

FIGURE D-14 Totals cells completed

The last formula to enter is for cell M21. This is not a total, but an overall percentage of Vehicle Capacity Utilized for all the vehicles used. This calculation uses the same formula as the cell above it, so you can simply copy cell M20 to cell M21. The formula for this cell is =G21/L21, which is Volume Required divided by Total Vehicle Capacity, expressed as a percentage. Your completed spreadsheet should look like Figure D-15.

Distance/Demand Table			Store Demand			Vehicle Loading							Cost
Distance Table (from Memphis Plant)	Miles	EB	CT	TM	Volume Required	Trucks	Volume for Trucks	Tractor-Trailers	Volume for Tractor-Trailers	Total Vehicle Capacity	% of Vehicle Capacity Utilized	Shipping Cost	
Philadelphia Store	1010	140	96	86	5012	1	1500	1	2350	3850	130%	$2,323.00	
Atlanta Store	380	76	81	63	3513	1	1500	1	2350	3850	91%	$874.00	
Miami Store	1000	56	64	52	2776	1	1500	1	2350	3850	72%	$2,300.00	
Chicago Store	540	115	130	150	6630	1	1500	1	2350	3850	172%	$1,242.00	
Los Angeles Store	1810	150	135	180	7785	1	1500	1	2350	3850	202%	$4,163.00	
				Totals:	25716	5	7500	5	11750	19250	134%	$10,902.00	
Fill Legend:			Changing Cells									Total Cost	
			Optimization Cell										

FIGURE D-15 Completed Calculations and Results section

Working the Model Manually

Now that you have a working model, you could manipulate the number of trucks and tractor-trailers manually to obtain a solution to the shipping problem. You would need to observe the following rules (or constraints):

1. Assign enough Total Vehicle Capacity to meet the Volume Required for each destination. (In other words, you cannot exceed 100% of Vehicle Capacity Utilized.)
2. The total number of trucks and tractor-trailers you assign cannot exceed the number available in your fleet.

Try to assign your trucks and tractor-trailers to meet your shipping requirements, and note the total shipping costs—you may get lucky and come up with an optimal solution. The tractor-trailers are more cost efficient than the trucks, but the problem is complicated by the fact that you want to achieve the best capacity utilization as well. In some instances, the trucks may be a better fit. Figure D-16 shows a sample solution determined from working the problem manually.

FIGURE D-16   Manual attempt to solve the vehicle loading problem optimally

This probably looks like a good solution—after all, you have not violated any of your constraints, and you have a 94% average vehicle capacity utilization. But is it the most cost-effective solution for your company? This is where Solver comes in.

Setting Up Solver Using the Solver Parameters Window

To access the Solver pane, click the Data tab on the Ribbon, then click Solver in the Analysis group on the far right side of the Ribbon. The Solver Parameters window appears (see Figure D-17).

> **NOTE**
>
> Solver in Excel 2010 has changed significantly from earlier versions of Excel. It allows three different calculation methods, and it allows you to specify an amount of time and number of iterations to perform before Excel ends the calculation. Refer to Microsoft Help for more information.

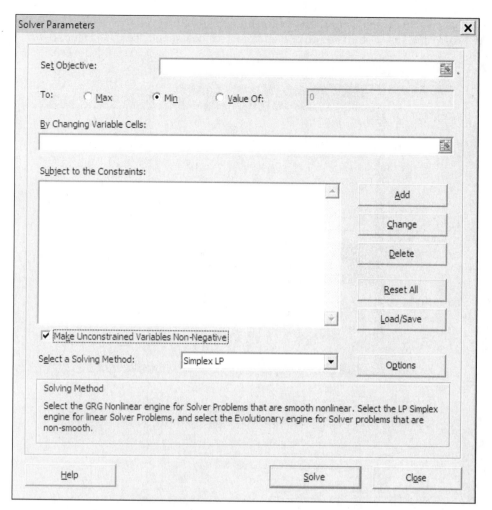

FIGURE D-17 Solver Parameters window

The Solver Parameters window in Excel 2010 looks intimidating at first. However, to solve linear optimization problems, you have to satisfy only three sets of conditions by filling in the following fields:

- Set Objective—Specify the optimization cell.
- By Changing Variable Cells—Specify the changing cells in your worksheet.
- Subject to the Constraints—Define all of the conditions and limitations that must be met when seeking the optimal solution.

The following sections explain these fields in detail. You may also need to click the Options button and select one or more options for solving the problem. Most of the cases in this book are linear problems, so you can set the solving method to Simplex LP, as shown in Figure D-17. If this method does not work in later cases, you can select the GRG Nonlinear or Evolutionary method to try to solve the problem. Note that solving methods are available only in Excel 2010.

Optimization Cell and Changing Cells

To use Solver successfully, you must first specify the cell you want to optimize—in this case, the total shipping cost, or cell N21. To fill the Set Objective field, click the button at the right edge of the field, and then click cell N21 in the spreadsheet. You could also type the cell address in the window, but selecting the cell in the spreadsheet reduces your chance of entering the wrong cell address. Next, specify whether you want Solver to seek the maximum or minimum value for cell N21. Because you want to minimize the total shipping cost, click the radio button next to Min.

Next, tell Solver which cell values it will change to determine the optimal solution. Use the By Changing Variable Cells field to specify the range of cells that you want Solver to manipulate. Again, click the button at the right edge of the field, select the cells that contain the numbers of trucks (H16 to H20), and then hold down the

Ctrl key and select the cells that contain the numbers of tractor-trailers (J16 to J20). If you used a fill color for the changing cells, they will be easy to find and select. The Solver Parameters window should look like Figure D-18.

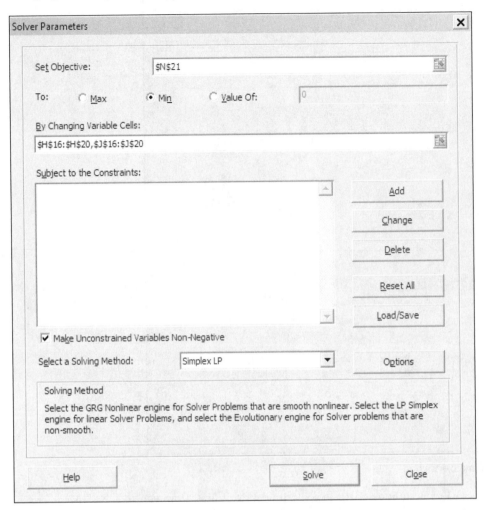

FIGURE D-18   Solver Parameters window with the objective cell and changing cells entered

Note that Solver has added absolute cell references (the $ signs before the column and row designators) for the cells you have specified. Solver will also add these references to the constraints you define. Solver adds the references to preserve the links to the cells in case you revise the worksheet in the future. In fact, you will make changes to the worksheet later in the tutorial.

Defining and Entering Constraints

For Solver to successfully determine the optimum solution for the shipping problem, you need to specify what constraints or rules it must observe to calculate the solution. Without constraints, Solver theoretically might calculate that the best solution is not to ship anything, resulting in a cost of zero. Furthermore, if you failed to define variables as positive numbers, Solver would select "negative trucks" to maximize "negative costs." Finally, the vehicles are indivisible units—you cannot assign a fraction of a vehicle for a fraction of the cost, so you must define your changing cells as integers to satisfy this constraint.

Aside from the preceding logical constraints, you have operational constraints as well. You cannot assign more vehicles than you have in your fleet, and the vehicles you assign must have at least as much total capacity as your shipping volume.

Before entering the constraints in the Solver Parameters window, it is a good idea to write them down in regular language. You must enter the following constraints for this model:

- All trucks and tractor-trailers in the changing cells must be integers greater than or equal to zero.
- The sums of trucks and tractor-trailers assigned (cells H21 and J21) must be less than or equal to the available trucks and tractor-trailers (cells F5 and F6, respectively).

- The Total Vehicle Capacity for the vehicles assigned to each store (cells L16 to L20) must be greater than or equal to the Volume Required to be shipped to each store (cells G16 to G20, respectively).

You are ready to enter the constraints as equations or inequalities in the Add Constraint window. To begin, click the Add button in the Solver Parameters window. In the window that appears (see Figure D-19), click the button at the right edge of the Cell Reference box, select cells H16 to H20, and then click the button again. Next, click the drop-down menu in the middle field and select > =. Then go to the Constraint field and type 0. Finally, click Add; otherwise, the constraint you defined will not be added to the list defined in the Solver Parameters window.

Add Constraint

Cell Reference:			Constraint:	
$H$16:$H$20		>=	0	

OK   Add   Cancel

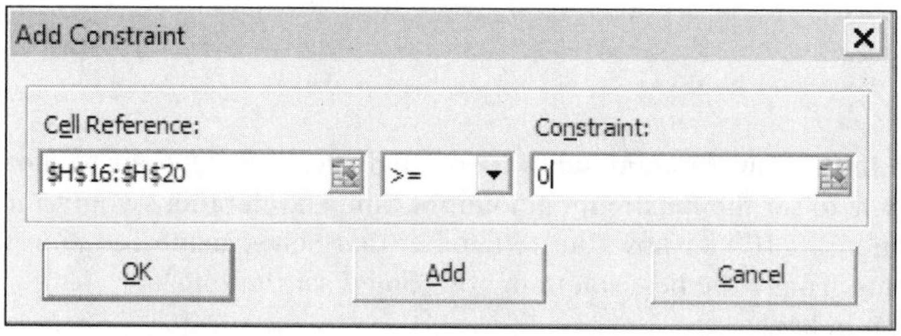

FIGURE D-19   Add Constraint window

You can continue to add constraints in the Add Constraint window. For this example, enter the constraints shown in the completed Solver Parameters window in Figure D-20. When you finish, click Add to save the last constraint, then click Cancel in the Add Constraint window to return to the Solver Parameters window.

Solver Parameters

Set Objective: $N$21

To: ◯ Max   ⦿ Min   ◯ Value Of: 0

By Changing Variable Cells:

$H$16:$H$20,$J$16:$J$20

Subject to the Constraints:

```
$H$16:$H$20 = integer
$H$16:$H$20 >= 0
$H$21 <= $F$5
$J$16:$J$20 = integer
$J$16:$J$20 >= 0
$J$21 <= $F$6
$L$16 >= $G$16
$L$17 >= $G$17
$L$18 >= $G$18
$L$19 >= $G$19
$L$20 >= $G$20
```

Add
Change
Delete
Reset All
Load/Save

☑ Make Unconstrained Variables Non-Negative

Select a Solving Method: Simplex LP   Options

Solving Method

Select the GRG Nonlinear engine for Solver Problems that are smooth nonlinear. Select the LP Simplex engine for linear Solver Problems, and select the Evolutionary engine for Solver problems that are non-smooth.

Help   Solve   Close

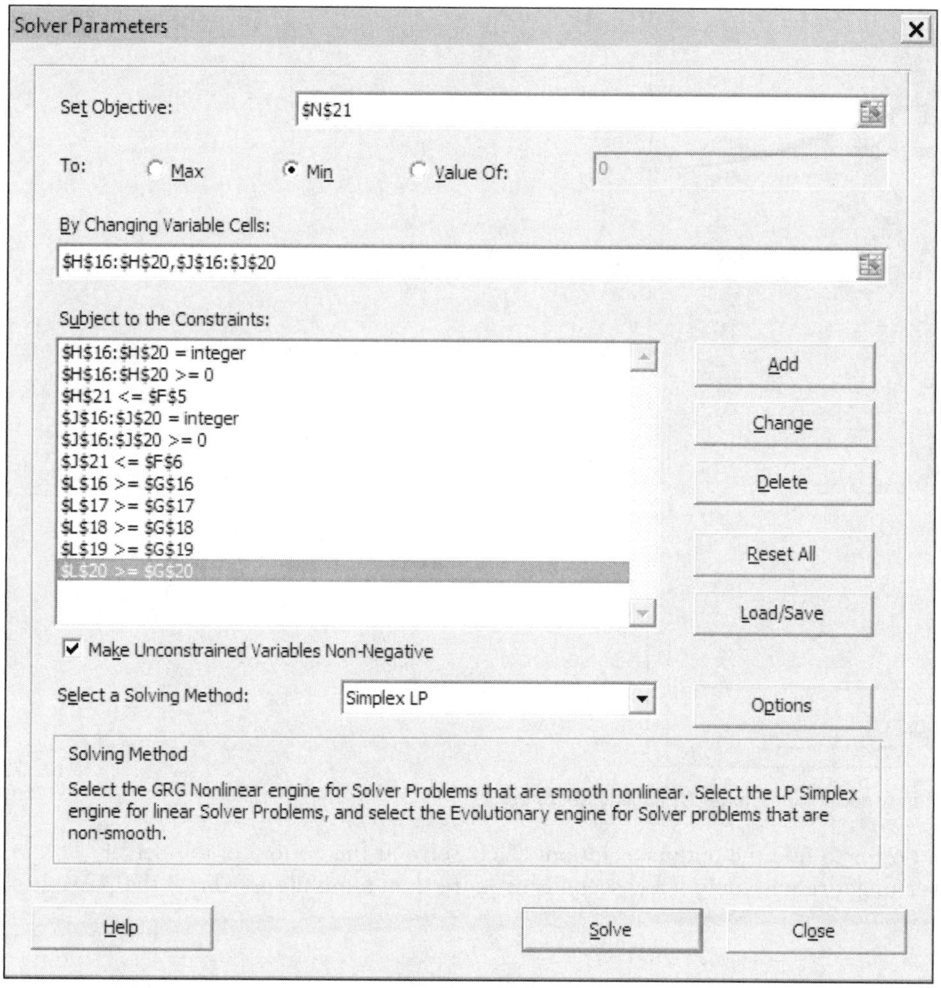

FIGURE D-20   Completed Solver Parameters window

If you have difficulty reading the constraints listed in Figure D-20, use the following list instead:

- $H$16:$H$20 = integer
- $H$16:$H$20 >= 0
- $H$21 <= $F$5
- $J$16:$J$20 = integer
- $J$16:$J$20 >= 0
- $J$21 <= $F$6
- $L$16 >= $G$16
- $L$17 >= $G$17
- $L$18 >= $G$18
- $L$19 >= $G$19
- $L$20 >= $G$20

You should also click the Options button in the Solver Parameters window and check the Options window shown in Figure D-21. You can use this window to set the maximum amount of time and iterations you want Solver to run before stopping. Leave both options at 100 for now, but remember that Solver may need more time and iterations for more complex problems. To get the best solution, you should set the Integer Optimality (%) to zero. Click OK to close the window.

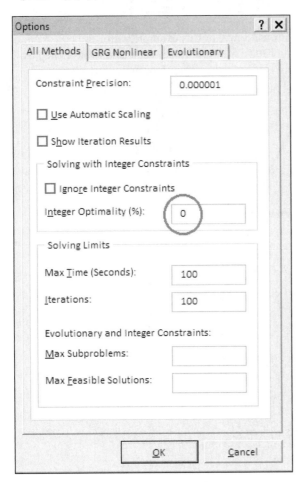

FIGURE D-21   Solver Options window with Integer Optimality set to zero

You are ready to run Solver to find the optimal solution. Click Solve at the bottom of the Solver Parameters window. Solver might require only a few seconds or more than a minute to run all the possible iterations—the status bar at the bottom of the Excel window displays iterations and possible solutions continuously until Solver finds an optimal solution or runs out of time (see Figure D-22).

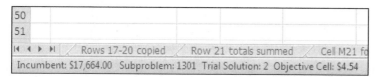

50					
51					
⏮ ◀ ▶ ⏭	Rows 17-20 copied	Row 21 totals summed	Cell M21 fo
Incumbent: $17,664.00 Subproblem: 1301 Trial Solution: 2 Objective Cell: $4,54

FIGURE D-22   Excel status bar showing Solver running through possible solutions

A new window will appear eventually, indicating that Solver has found an optimal solution to the problem (see Figure D-23). The portion of the spreadsheet that displays the assigned vehicles and shipping cost should be visible below the Solver Results window. Solver has assigned nine of the 12 trucks and all six tractor-trailers, for a total shipping cost of $17,398. The earlier manual attempt to solve the problem (see Figure D-16) assigned all 12 trucks and four tractor-trailers, for a total shipping cost of $18,122. Using Solver in this situation saved your company $724.

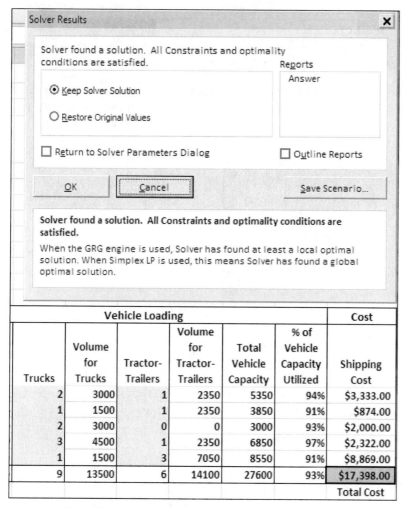

Trucks	Volume for Trucks	Tractor-Trailers	Volume for Tractor-Trailers	Total Vehicle Capacity	% of Vehicle Capacity Utilized	Shipping Cost
2	3000	1	2350	5350	94%	$3,333.00
1	1500	1	2350	3850	91%	$874.00
2	3000	0	0	3000	93%	$2,000.00
3	4500	1	2350	6850	97%	$2,322.00
1	1500	3	7050	8550	91%	$8,869.00
9	13500	6	14100	27600	93%	$17,398.00

FIGURE D-23   Solver Results window

If the Solver Results window does not report an optimal solution to the problem, it will report that the problem could not be solved given the changing cells and constraints you specified. For instance, if you had not had enough vehicles in your fleet to carry the required shipping volume to all the destinations, the

Solver Results window might have looked like Figure D-24. In the figure, your vehicle fleet was reduced to 10 trucks and five tractor-trailers, so Solver could not find a solution that satisfied the shipping volume constraints.

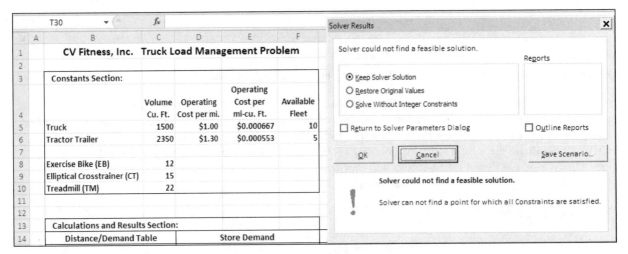

FIGURE D-24   Solver could not find a feasible solution with a reduced vehicle fleet

Fortunately, Solver did find an optimal solution. To update the spreadsheet with the new optimal values for the changing cells and optimization cell, click OK in the Solver Results window. You can also create an Answer Report by clicking the Answer option in the Solver Results window (see Figure D-25) and then clicking OK.

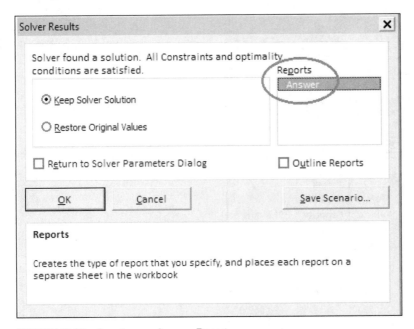

FIGURE D-25   Creating an Answer Report

Excel will create a report in a separate sheet called Answer Report 1. The Answer Report is shown in Figures D-26 and D-27.

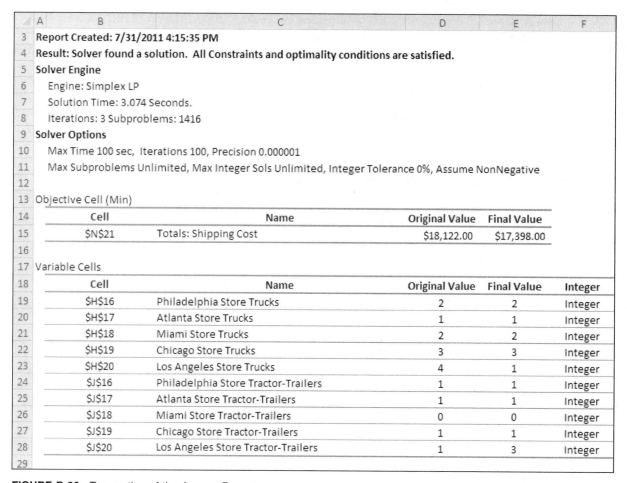

	A	B	C	D	E	F
3	Report Created: 7/31/2011 4:15:35 PM					
4	Result: Solver found a solution. All Constraints and optimality conditions are satisfied.					
5	Solver Engine					
6	Engine: Simplex LP					
7	Solution Time: 3.074 Seconds.					
8	Iterations: 3 Subproblems: 1416					
9	Solver Options					
10	Max Time 100 sec, Iterations 100, Precision 0.000001					
11	Max Subproblems Unlimited, Max Integer Sols Unlimited, Integer Tolerance 0%, Assume NonNegative					
12						
13	Objective Cell (Min)					
14		Cell	Name	Original Value	Final Value	
15		$N$21	Totals: Shipping Cost	$18,122.00	$17,398.00	
16						
17	Variable Cells					
18		Cell	Name	Original Value	Final Value	Integer
19		$H$16	Philadelphia Store Trucks	2	2	Integer
20		$H$17	Atlanta Store Trucks	1	1	Integer
21		$H$18	Miami Store Trucks	2	2	Integer
22		$H$19	Chicago Store Trucks	3	3	Integer
23		$H$20	Los Angeles Store Trucks	4	1	Integer
24		$J$16	Philadelphia Store Tractor-Trailers	1	1	Integer
25		$J$17	Atlanta Store Tractor-Trailers	1	1	Integer
26		$J$18	Miami Store Tractor-Trailers	0	0	Integer
27		$J$19	Chicago Store Tractor-Trailers	1	1	Integer
28		$J$20	Los Angeles Store Tractor-Trailers	1	3	Integer
29						

FIGURE D-26   Top portion of the Answer Report

	A	B	C	D	E	F	G
31	Constraints						
32		Cell	Name	Cell Value	Formula	Status	Slack
33	$H$21		Totals: Trucks	9	$H$21<=$F$5	Not Binding	3
34	$J$21		Totals: Tractor-Trailers	6	$J$21<=$F$6	Binding	0
35	$L$16		Philadelphia Store Total Vehicle Capacity	5350	$L$16>=$G$16	Not Binding	338
36	$L$17		Atlanta Store Total Vehicle Capacity	3850	$L$17>=$G$17	Not Binding	337
37	$L$18		Miami Store Total Vehicle Capacity	3000	$L$18>=$G$18	Not Binding	224
38	$L$19		Chicago Store Total Vehicle Capacity	6850	$L$19>=$G$19	Not Binding	220
39	$L$20		Los Angeles Store Total Vehicle Capacity	8550	$L$20>=$G$20	Not Binding	765
40	$H$16		Philadelphia Store Trucks	2	$H$16>=0	Binding	0
41	$H$17		Atlanta Store Trucks	1	$H$17>=0	Binding	0
42	$H$18		Miami Store Trucks	2	$H$18>=0	Binding	0
43	$H$19		Chicago Store Trucks	3	$H$19>=0	Binding	0
44	$H$20		Los Angeles Store Trucks	1	$H$20>=0	Binding	0
45	$J$16		Philadelphia Store Tractor-Trailers	1	$J$16>=0	Binding	0
46	$J$17		Atlanta Store Tractor-Trailers	1	$J$17>=0	Binding	0
47	$J$18		Miami Store Tractor-Trailers	0	$J$18>=0	Binding	0
48	$J$19		Chicago Store Tractor-Trailers	1	$J$19>=0	Binding	0
49	$J$20		Los Angeles Store Tractor-Trailers	3	$J$20>=0	Binding	0
50	$H$16:$H$20=Integer						
51	$J$16:$J$20=Integer						
52							
53							

Manual Solution   Solver Setup and Run   **Answer Report 1**   Solution 1

FIGURE D-27   Bottom portion of the Answer Report—note the new tab created by Solver

The Answer Report gives you a wealth of information about the solution. The top portion displays the original and final values of the Objective cell. The second part of the report displays the original and final values of the changing cells. The last part of the report lists the constraints. Binding constraints are those that reached their maximum or minimum value; nonbinding constraints did not.

Perhaps a savings of $724 does not seem significant—however, this problem does not have a specified time frame. The example probably represents one week of shipments for CV Fitness. The store demands will change from week to week, but you could use Solver each time to optimize the truck assignments. In a 50-week business year, the savings that Solver helps you find in shipping costs could be well over $30,000!

Go to the File tab to print the worksheets you created. Save the Excel file as **CV Fitness Trucking Problem.xlsx**, then select the Save As command in the File tab to create a new file called **CV Fitness Trucking Problem 2.xlsx**. You will use the new file in the next section.

EXTENDING THE EXAMPLE

Like all successful companies, CV Fitness looks for ways to grow its business and optimize its costs. Your management team is considering two changes:

- Opening two new stores and expanding the vehicle fleet if necessary
- Improving product design and packaging to reduce the shipping volume of the treadmill from 22 cubic feet to 17 cubic feet

You have been asked to modify your model to see the new requirements for each change separately. The two new stores would be in Denver and Phoenix, and they are 1,040 and 1,470 miles from the Memphis plant, respectively. If necessary, open the CV Fitness Trucking Problem 2.xlsx file, then right-click row 21 at the left worksheet border. Click Insert to enter a new row between rows 20 and 21. Repeat the steps to insert a second new row. Your spreadsheet should look like Figure D-28. Do not worry about the borders for now—you can fix them later.

	A	B	C	D	E	F	G	H	I	J	K	L	M	N
13		Calculations and Results Section:												
14		Distance/Demand Table			Store Demand				Vehicle Loading					Cost
15		Distance Table (from Memphis Plant)	Miles	EB	CT	TM	Volume Required	Trucks	Volume for Trucks	Tractor-Trailers	Volume for Tractor-Trailers	Total Vehicle Capacity	% of Vehicle Capacity Utilized	Shipping Cost
16		Philadelphia Store	1010	140	96	86	5012	2	3000	1	2350	5350	94%	$3,333.00
17		Atlanta Store	380	76	81	63	3513	1	1500	1	2350	3850	91%	$874.00
18		Miami Store	1000	56	64	52	2776	2	3000	0	0	3000	93%	$2,000.00
19		Chicago Store	540	115	130	150	6630	3	4500	1	2350	6850	97%	$2,322.00
20		Los Angeles Store	1810	150	135	180	7785	1	1500	3	7050	8550	91%	$8,869.00
21														
22														
23						Totals:	25716	9	13500	6	14100	27600	93%	$17,398.00
24		Fill Legend:			Changing Cells									Total Cost
25					Optimization Cell									

FIGURE D-28   Distance/Demand table with two blank rows inserted for the new stores

Enter the two new stores in cells B21 and B22, enter their distances in cells C21 and C22, and enter the Store Demands in cells D21 through F22, as shown in Figure D-29. When you complete this part of the table, fix the borders to include the two new stores. Select the area in the table you want to fix, click the No Borders button to clear the old borders, highlight the area to which you want to add the border, and then click the Outside Borders button.

	A	B	C	D	E	F
13		Calculations and Results Section:				
14		Distance/Demand Table			Store Demand	
15		Distance Table (from Memphis Plant)	Miles	EB	CT	TM
16		Philadelphia Store	1010	140	96	86
17		Atlanta Store	380	76	81	63
18		Miami Store	1000	56	64	52
19		Chicago Store	540	115	130	150
20		Los Angeles Store	1810	150	135	180
21		Denver Store	1040	74	67	43
22		Phoenix Store	1470	41	28	37
23						Totals:
24		Fill Legend:		Changing Cells		
25				Optimization Cell		

FIGURE D-29 Distance/Demand table with new store locations and demands entered

Next, copy the formulas from cells G20 to N20 to the two new rows in the Vehicle Loading and Cost sections of the table. Select cells G20 to N20, right-click, and click Copy on the menu. Then select cells G21 to N22 and click Paste in the Clipboard group. Your table should look like Figure D-30.

	G	H	I	J	K	L	M	N
13								
14			Vehicle Loading					Cost
15	Volume Required	Trucks	Volume for Trucks	Tractor-Trailers	Volume for Tractor-Trailers	Total Vehicle Capacity	% of Vehicle Capacity Utilized	Shipping Cost
16	5012	2	3000	1	2350	5350	94%	$3,333.00
17	3513	1	1500	1	2350	3850	91%	$874.00
18	2776	2	3000	0	0	3000	93%	$2,000.00
19	6630	3	4500	1	2350	6850	97%	$2,322.00
20	7785	1	1500	3	7050	8550	91%	$8,869.00
21	2839	1	1500	3	7050	8550	33%	$5,096.00
22	1726	1	1500	3	7050	8550	20%	$7,203.00
23	30281	9	13500	6	14100	27600	110%	$17,398.00
24								Total Cost

FIGURE D-30 Formulas from row 20 copied into rows 21 and 22

Note that most cells in the Totals row have not changed—their formulas need to be updated to include the values in rows 21 and 22. To quickly check which cells you need to update, display the formulas in the Totals row. Hold down the Ctrl key and press the ~ key (on most keyboards, this key is next to the "1" key). The Vehicle Loading and Cost sections now display formulas in the cells (see Figure D-31).

	G	H	I	J	K	L	M	N
13								
14				Vehicle Loading				Cost
15	Volume Required	Trucks	Volume for Trucks	Tractor- Trailers	Volume for Tractor- Trailers	Total Vehicle Capacity	% of Vehicle Capacity Utilized	Shipping Cost
16	=D16*$C$8+E16*$C$9+F16*$C$10	2	=H16*$C$5	1	=J16*$C$6	=I16+K16	=G16/L16	=H16*C16*$D$5+J16*C16*$D$6
17	=D17*$C$8+E17*$C$9+F17*$C$10	1	=H17*$C$5	1	=J17*$C$6	=I17+K17	=G17/L17	=H17*C17*$D$5+J17*C17*$D$6
18	=D18*$C$8+E18*$C$9+F18*$C$10	2	=H18*$C$5	0	=J18*$C$6	=I18+K18	=G18/L18	=H18*C18*$D$5+J18*C18*$D$6
19	=D19*$C$8+E19*$C$9+F19*$C$10	3	=H19*$C$5	1	=J19*$C$6	=I19+K19	=G19/L19	=H19*C19*$D$5+J19*C19*$D$6
20	=D20*$C$8+E20*$C$9+F20*$C$10	1	=H20*$C$5	3	=J20*$C$6	=I20+K20	=G20/L20	=H20*C20*$D$5+J20*C20*$D$6
21	=D21*$C$8+E21*$C$9+F21*$C$10	1	=H21*$C$5	3	=J21*$C$6	=I21+K21	=G21/L21	=H21*C21*$D$5+J21*C21*$D$6
22	=D22*$C$8+E22*$C$9+F22*$C$10	1	=H22*$C$5	3	=J22*$C$6	=I22+K22	=G22/L22	=H22*C22*$D$5+J22*C22*$D$6
23	=SUM(G16:G22)	=SUM(H16:H20)	=SUM(I16:I20)	=SUM(J16:J20)	=SUM(K16:K20)	=SUM(L16:L20)	=G23/L23	=SUM(N16:N20)
24								Total Cost

FIGURE D-31   Vehicle Loading and Cost sections with formulas displayed in the cells

You must update any Totals cells that do not include the contents of rows 21 and 22. For example, you need to update the Totals cells H23 through L23 and cell N23. Cell M23 is not really a total; it is a cumulative ratio formula, so you do not need to update the cell. Use the following formulas to revise the Totals cells:

- Cell H23: =SUM(H16:H22)
- Cell I23: =SUM(I16:I22)
- Cell J23: =SUM(J16:J22)
- Cell K23: =SUM(K16:K22)
- Cell L23: =SUM(L16:L22)
- Cell N23: =SUM(N16:N22)

The updated sections should look like Figure D-32.

	G	H	I	J	K	L	M	N
14				Vehicle Loading				Cost
15	Volume Required	Trucks	Volume for Trucks	Tractor- Trailers	Volume for Tractor- Trailers	Total Vehicle Capacity	% of Vehicle Capacity Utilized	Shipping Cost
16	5012	2	3000	1	2350	5350	94%	$3,333.00
17	3513	1	1500	1	2350	3850	91%	$874.00
18	2776	2	3000	0	0	3000	93%	$2,000.00
19	6630	3	4500	1	2350	6850	97%	$2,322.00
20	7785	1	1500	3	7050	8550	91%	$8,869.00
21	2839	1	1500	3	7050	8550	33%	$5,096.00
22	1726	1	1500	3	7050	8550	20%	$7,203.00
23	30281	11	16500	12	28200	44700	68%	$29,697.00
24								Total Cost

FIGURE D-32   Vehicle Loading and Cost sections with the formulas updated

You are ready to use Solver to determine the optimal vehicle assignment. Click Solver in the Analysis group of the Data tab. You should notice immediately that you must revise the changing cells to include the two new stores; you must also change some of the constraints and add others. Solver has already updated the Objective cell from $N$21 to $N$23 and has updated the $H$23<=$F$5 and $J$23<=$F$6 constraints for vehicle fleet size. To update the changing cells, click the button to the right of the By Changing Variable Cells field and select the cells again, or edit the formula in the window by changing cell address $H$20 to $H$22 and cell address $J$20 to $J$22.

To change a constraint, select the one you want to change, and then click Change (see Figure D-33).

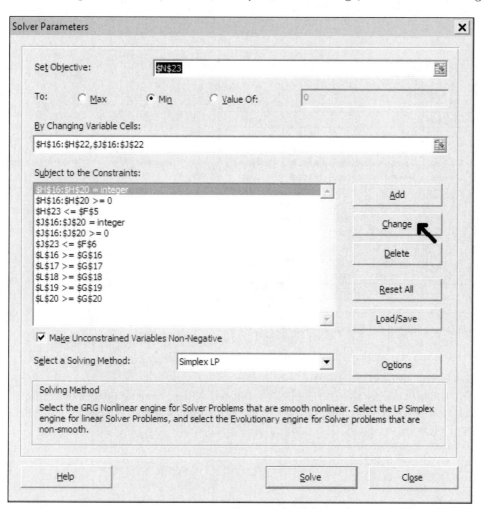

FIGURE D-33 Selecting a constraint to change

When you click Change, the Change Constraint window appears. Click the Cell Reference button; the selected cells will appear on the spreadsheet with a moving marquee around them (see Figure D-34). Highlight the new group of cells; when the new range appears in the Cell Reference field, click OK. The Solver Parameters window appears with the constraint changed.

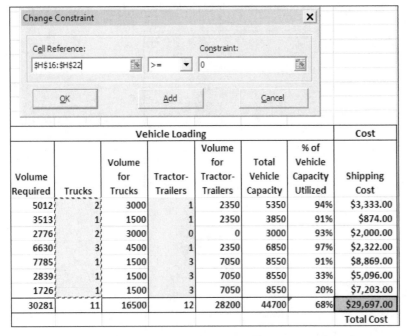

FIGURE D-34 Adding cells H21 and H22 to the Trucks constraint cell range

You also need to update or add the following constraints:

- Update $J$16:$J$20 >=0 to $J$16:$J$22 >=0.
- Update $H$16:$H$20 = integer to $H$16:$H$22 = integer. When changing integer constraints, you must click "int" in the middle field of the Change Constraint window; otherwise, you will receive an error message.
- Update $J$16:$J$20 = integer to $J$16:$J$22 = integer.
- Add constraint $L$21 >= $G$21 (see Figure D-35).
- Add constraint $L$22 >= $G$22.

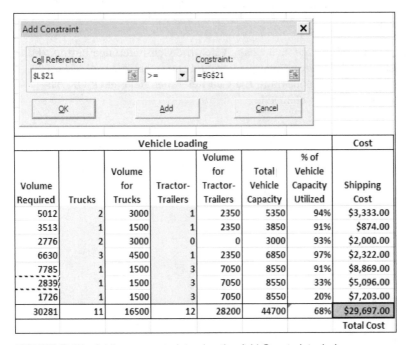

FIGURE D-35 Adding a constraint using the Add Constraint window

You are ready to solve the shipping problem to include the new stores in Denver and Phoenix. Figure D-36 shows the updated Solver Parameters window.

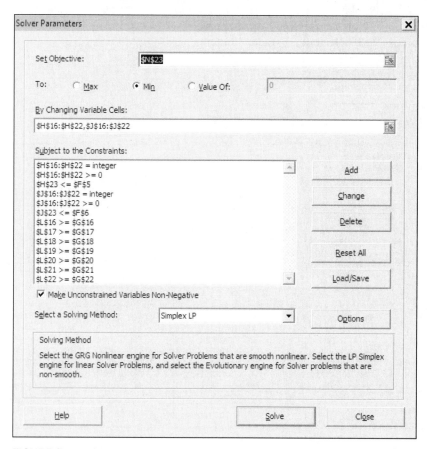

FIGURE D-36   Solver parameters updated for shipping to seven stores

Before you run Solver again, you might want to attempt to assign the vehicles manually, because your fleet may not be large enough to handle two more stores. In this case, you will quickly realize that the vehicle fleet is at least one truck or tractor-trailer short of the minimum required to ship the needed volume. You can confirm this by running Solver (see Figure D-37).

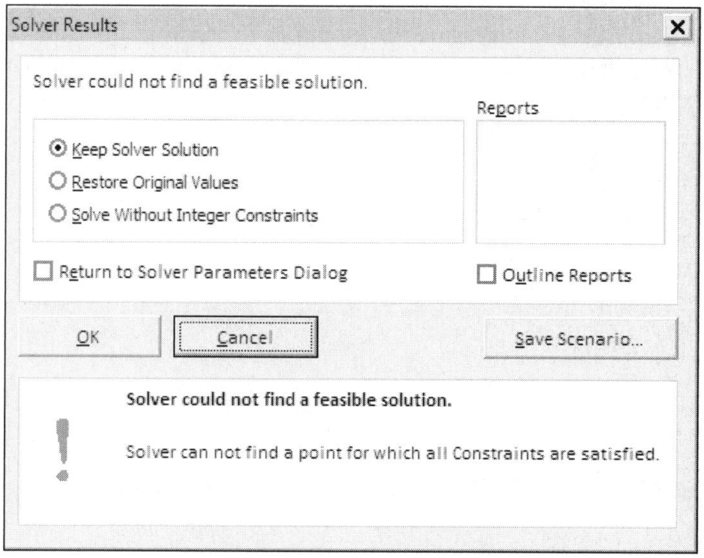

FIGURE D-37   Vehicle fleet does not meet minimum requirements

The Solver Results window confirms that your truck fleet is too small, so change the value in cell F5 from 12 to 13 to add another truck to your fleet, and then run Solver again. As you add more stores and vehicles to make the problem more complex, Solver will take longer to run, especially on older computers. You may have to wait a minute or more for Solver to finish its iterations and find an answer (see Figure D-38). In this example, Solver recommends that you use 13 trucks and six tractor-trailers.

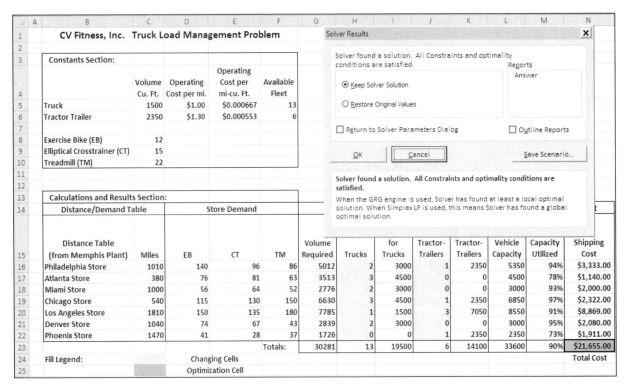

FIGURE D-38   Solver's solution

Select Answer in the Reports list to add an Answer Report to the workbook, and then click OK. You can keep or delete the old Answer Report 1 tab from the earlier workbook. The new Answer Report is in a new worksheet named Answer Report 2.

You can meet the shipping requirements by adding one more truck, but is it really the most cost-effective solution? What if you add a tractor-trailer instead? Set the number of trucks back to 12, and add a tractor-trailer by entering 7 instead of 6 in cell F6. Run Solver again.

This time Solver finds a less expensive solution, as shown in Figures D-39 and D-40. At first it does not make sense—how can adding a more expensive vehicle (a tractor-trailer) reduce the overall expense? In fact, the additional tractor-trailer has replaced two trucks. With seven tractor-trailers, you only need 11 trucks instead of the original 13.

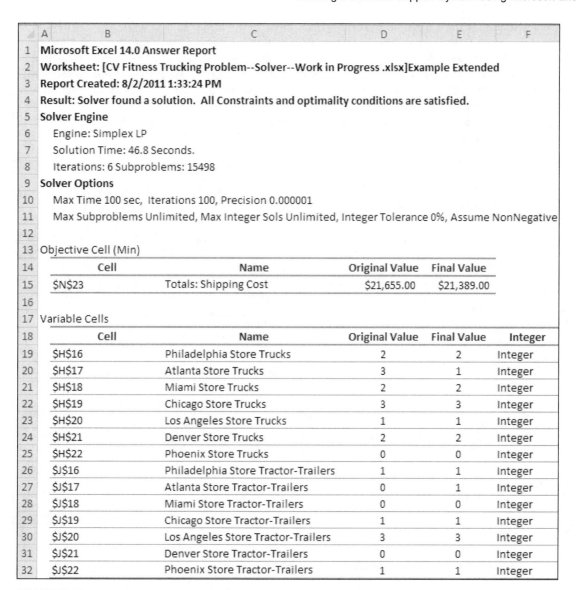

	A	B	C	D	E	F
1	Microsoft Excel 14.0 Answer Report					
2	Worksheet: [CV Fitness Trucking Problem--Solver--Work in Progress .xlsx]Example Extended					
3	Report Created: 8/2/2011 1:33:24 PM					
4	Result: Solver found a solution. All Constraints and optimality conditions are satisfied.					
5	Solver Engine					
6	Engine: Simplex LP					
7	Solution Time: 46.8 Seconds.					
8	Iterations: 6 Subproblems: 15498					
9	Solver Options					
10	Max Time 100 sec, Iterations 100, Precision 0.000001					
11	Max Subproblems Unlimited, Max Integer Sols Unlimited, Integer Tolerance 0%, Assume NonNegative					
12						
13	Objective Cell (Min)					
14		Cell	Name	Original Value	Final Value	
15	$N$23		Totals: Shipping Cost	$21,655.00	$21,389.00	
16						
17	Variable Cells					
18		Cell	Name	Original Value	Final Value	Integer
19	$H$16		Philadelphia Store Trucks	2	2	Integer
20	$H$17		Atlanta Store Trucks	3	1	Integer
21	$H$18		Miami Store Trucks	2	2	Integer
22	$H$19		Chicago Store Trucks	3	3	Integer
23	$H$20		Los Angeles Store Trucks	1	1	Integer
24	$H$21		Denver Store Trucks	2	2	Integer
25	$H$22		Phoenix Store Trucks	0	0	Integer
26	$J$16		Philadelphia Store Tractor-Trailers	1	1	Integer
27	$J$17		Atlanta Store Tractor-Trailers	0	1	Integer
28	$J$18		Miami Store Tractor-Trailers	0	0	Integer
29	$J$19		Chicago Store Tractor-Trailers	1	1	Integer
30	$J$20		Los Angeles Store Tractor-Trailers	3	3	Integer
31	$J$21		Denver Store Tractor-Trailers	0	0	Integer
32	$J$22		Phoenix Store Tractor-Trailers	1	1	Integer

FIGURE D-39   Answer Report 3 displays a more cost-effective solution

	G	H	I	J	K	L	M	N
13								
14	Vehicle Loading							Cost
15	Volume Required	Trucks	Volume for Trucks	Tractor-Trailers	Volume for Tractor-Trailers	Total Vehicle Capacity	% of Vehicle Capacity Utilized	Shipping Cost
16	5012	2	3000	1	2350	5350	94%	$3,333.00
17	3513	1	1500	1	2350	3850	91%	$874.00
18	2776	2	3000	0	0	3000	93%	$2,000.00
19	6630	3	4500	1	2350	6850	97%	$2,322.00
20	7785	1	1500	3	7050	8550	91%	$8,869.00
21	2839	2	3000	0	0	3000	95%	$2,080.00
22	1726	0	0	1	2350	2350	73%	$1,911.00
23	30281	11	16500	7	16450	32950	92%	$21,389.00
24								Total Cost

FIGURE D-40   Seven tractor-trailers and 11 trucks are the optimal mix

You have a solution for the expansion to seven stores. Save your workbook, and then create a new workbook using the Save As command. Name the new workbook **CV Fitness Trucking Problem 3.xlsx**.

Next, evaluate the potential cost savings if the company redesigns its treadmill product and packaging to reduce the shipping volume from 22 cubic feet to 17 cubic feet. Your engineers report that the redesign will cost approximately $10,000. If you can save at least $500 per shipment, the project will pay for itself in less than six months (20 weekly shipments).

Go to cell C10 on the worksheet, replace 22 with 17, and run Solver again. When Solver finds the solution, select Answer to create another Answer Report, and then click OK. See Figure D-41.

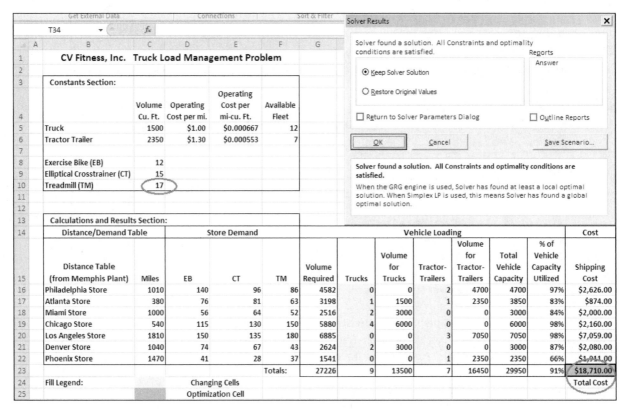

FIGURE D-41   Solver solution with redesigned treadmill and packaging

Check the Answer Report to see the cost difference between shipping the old treadmills and the redesigned models (see Figure D-42). The cost savings for one shipment is $2,679, which is more than five times the minimum savings you needed. You should go ahead with the project.

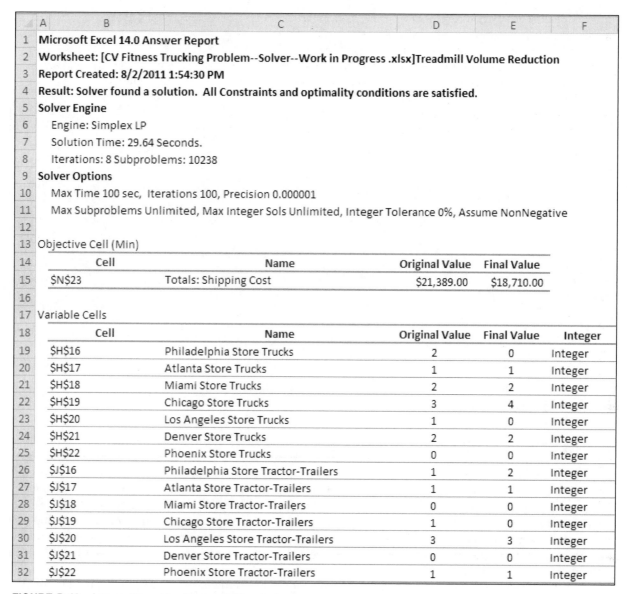

	A	B	C	D	E	F
1	**Microsoft Excel 14.0 Answer Report**					
2	**Worksheet: [CV Fitness Trucking Problem--Solver--Work in Progress .xlsx]Treadmill Volume Reduction**					
3	**Report Created: 8/2/2011 1:54:30 PM**					
4	**Result: Solver found a solution. All Constraints and optimality conditions are satisfied.**					
5	**Solver Engine**					
6	Engine: Simplex LP					
7	Solution Time: 29.64 Seconds.					
8	Iterations: 8 Subproblems: 10238					
9	**Solver Options**					
10	Max Time 100 sec, Iterations 100, Precision 0.000001					
11	Max Subproblems Unlimited, Max Integer Sols Unlimited, Integer Tolerance 0%, Assume NonNegative					
12						
13	Objective Cell (Min)					
14	Cell		Name	Original Value	Final Value	
15	$N$23		Totals: Shipping Cost	$21,389.00	$18,710.00	
16						
17	Variable Cells					
18	Cell		Name	Original Value	Final Value	Integer
19	$H$16		Philadelphia Store Trucks	2	0	Integer
20	$H$17		Atlanta Store Trucks	1	1	Integer
21	$H$18		Miami Store Trucks	2	2	Integer
22	$H$19		Chicago Store Trucks	3	4	Integer
23	$H$20		Los Angeles Store Trucks	1	0	Integer
24	$H$21		Denver Store Trucks	2	2	Integer
25	$H$22		Phoenix Store Trucks	0	0	Integer
26	$J$16		Philadelphia Store Tractor-Trailers	1	2	Integer
27	$J$17		Atlanta Store Tractor-Trailers	1	1	Integer
28	$J$18		Miami Store Tractor-Trailers	0	0	Integer
29	$J$19		Chicago Store Tractor-Trailers	1	0	Integer
30	$J$20		Los Angeles Store Tractor-Trailers	3	3	Integer
31	$J$21		Denver Store Tractor-Trailers	0	0	Integer
32	$J$22		Phoenix Store Tractor-Trailers	1	1	Integer

FIGURE D-42   Answer Report for the treadmill redesign

When you finish examining the Answer Report, save your file and then close it. To close the workbook, click the File tab and then click Close (see Figure D-43).

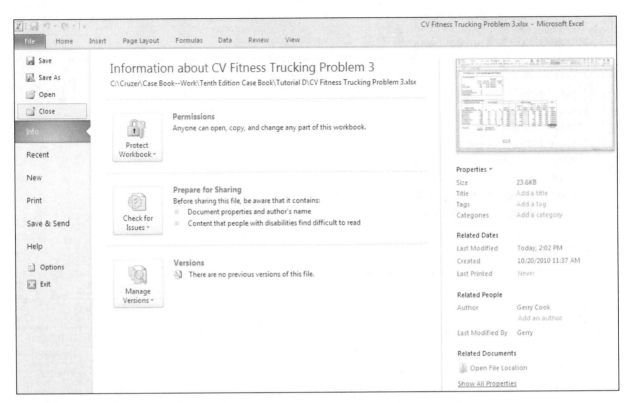

FIGURE D-43   Closing the Excel workbook

USING SOLVER ON A NEW PROBLEM

A common problem in manufacturing businesses is deciding on a product mix for different items in the same product family. Sensuous Scents Inc. makes a premium collection of perfume, cologne, and body spray for sale in large department stores and boutiques. The primary ingredient is ambergris, a valuable digestive excretion from whales that is harvested without harming the animals. Ambergris costs more than $9,000 per pound and is very difficult to obtain in large quantities; Sensuous Scents can obtain only about 20 pounds of ambergris each year. The other ingredients—deionized water, ethanol, and various additives—are available in unlimited quantities for a reasonable cost.

You have been asked to create a spreadsheet model for Solver to determine the optimal product mix that maximizes Sensuous Scents' net income after taxes.

Setting up the Spreadsheet

The sections in this spreadsheet are different from those in the preceding trucking problem. You will create a Constants section, a Bill of Materials section for the three products, a Quantity Manufactured section that contains the changing cells, a Calculations section (to calculate ambergris usage, manufacturing costs, and sales revenue per product line), and an Income Statement section to determine the net income after taxes, which will be the optimization cell.

AT THE KEYBOARD

Start a new file called **Sensuous Scents Inc.xlsx** and set up the spreadsheet.

Spreadsheet Title and Constants Section

Your spreadsheet title and Constants section should look like Figure D-44. A discussion of the section entries follows the figure.

			Body Spray	Cologne	Perfume
Sensuous Scents Inc. Product Mix					
Constants:			Body Spray	Cologne	Perfume
Sales Price per bottle			$11.95	$21.00	$53.00
Conversion Cost per Unit (Direct Labor plus Manufacturing Overhead)			$2.60	$6.50	$13.00
Minimum Sales Demand			60000	25000	12000
Income Tax Rate		0.32			
Sales, General and Administrative Expenses per Dollar Revenue		0.30			
Available Ambergris (lbs)		20			
Cost per lb, Deionized Water		$0.50			
Cost per lb, Ethanol		$1.00			
Cost per lb, other Additives		$182.00			
Cost per lb, Ambergris		$9,072.00			

FIGURE D-44 Spreadsheet title and Constants section for Sensuous Scents Inc.

- Sales Price per bottle—These values are the sales prices for each of the three products.
- Conversion Cost per Unit—These values are the direct labor costs plus the manufacturing overhead costs budgeted per unit manufactured. A conversion cost is often used in industries that manufacture liquid products.
- Minimum Sales Demand—These values reflect the forecast minimum sales demand that you must supply to your customers. These values will be used later as constraints.
- Income Tax Rate—The rate is 32% of your pretax income. No taxes are paid on losses.
- Sales, General and Administrative Expenses per Dollar Revenue—This value is an estimate of the non-manufacturing costs that Sensuous Scents will incur per dollar of sales revenue. These expenses are subtracted from the Gross Profit value in the Income Statement section to obtain Net Income before taxes.
- Available Ambergris (lbs.)—This value is the amount of ambergris that Sensuous Scents obtained this year for production.
- Cost per lb., Deionized Water—This value is the current cost per pound of deionized water.
- Cost per lb., Ethanol—This value is the current cost per pound of ethanol.
- Cost per lb., other Additives—Scent products contain other additives and fixatives to enhance or preserve the fragrance. This value is the cost per pound of the other additives.
- Cost per lb., Ambergris—This value is the current market price per pound of naturally harvested ambergris. Again, no whales are harmed to obtain the ambergris.

The rest of the cells are filled with a gray background to indicate that you will not use their values or formulas. The section is arranged this way to maintain one column per product all the way down the spreadsheet, which will simplify writing the formulas later.

Bill of Materials Section

Your spreadsheet should contain a Bill of Materials section, as shown in Figure D-45. The section entries are explained after the figure. A bill of materials is a list of raw materials and ingredients required to make one unit of a product.

				Body Spray	Cologne	Perfume
Bill of Materials:				Body Spray	Cologne	Perfume
Deionized Water (lb)				0.4	0.1	0.05
Ethanol (lb)				0.1	0.02	0.01
Other Additives (lb)				0.01	0.001	0.0001
Ambergris (lb)				0.0001	0.00018	0.00055

FIGURE D-45 Bill of Materials section

- Deionized Water (lb.)—The amount of deionized water required to make one unit of each product
- Ethanol (lb.)—The amount of ethanol required to make one unit of each product
- Other Additives (lb.)—The amount of other additives required to make one unit of each product
- Ambergris (lb.)—The amount of ambergris required to make one unit of each product

Extremely small quantities of ambergris and other additives are required to make one bottle of each product. Also, each product requires a different amount of ambergris. Check the values to make sure you entered the correct number of decimal places.

Quantity Manufactured (Changing Cells) Section

This model contains a separate Changing Cells section called Quantity Manufactured, as shown in Figure D-46. This section contains the cells that you want Solver to manipulate to achieve the highest net income after taxes.

	A	B	C	D	E	F
20						
21		Quantity Manufactured (Changing Cells)		Body Spray	Cologne	Perfume
22		Units Produced		60000	25000	12000

FIGURE D-46 Quantity Manufactured (changing cells) section

Cells D22, E22, and F22 are yellow to indicate that Solver will change them to reach an optimal solution. To begin, enter the minimum sales demand in these cells, which will remind you to specify the minimum demand constraints from the Constants section in the Solver Parameters window.

Calculations Section

Your model should contain the Calculations section shown in Figure D-47.

	A	B	C	D	E	F	G
23							
24		Calculations:		Body Spray	Cologne	Perfume	Totals
25		Lbs of Ambergris Used					
26		Manufacturing Cost per Unit (Materials Costs plus Conversion Cost)					
27		Total Manufacturing Costs per Product Line					
28		Sales Revenues per Product Line					

FIGURE D-47 Calculations section

The section contains the following calculations:

- Lbs. of Ambergris Used—This value is the pounds of ambergris per unit from the Bill of Materials section, multiplied by Units Produced from the Quantity Manufactured section for each of the three products. The Totals cell (G25) is the sum of cells D25, E25, and F25. Use the value in this cell to specify the constraint that you have only 20 pounds of ambergris available to use for raw materials (Constants section, cell C9).
- Manufacturing Cost per Unit (Materials Costs plus Conversion Cost)—To get this value, write a formula that multiplies the unit cost for each of the four product ingredients by the amount per unit specified in the bill of materials, multiplied by Units Produced. The total materials costs for the four ingredients are added together, and then the Conversion Cost per Unit is added from the Constants section to obtain the Manufacturing Cost per Unit. Enter the following formula for the Body Spray Manufacturing Cost per Unit in cell D26:

=$C$10*D16+$C$11*D17+$C$12*D18+$C$13*D19+D5

Use absolute cell references for the cells that hold values for costs per pound ($C$10, $C$11, $C$12, and $C$13). By doing so, you can copy the body spray formula to the Manufacturing Cost per Unit cells for the cologne and perfume values (cells E26 and F26). The Totals cell (G26) is not used in this row—you can fill the cell in gray to indicate that it is not used.
- Total Manufacturing Costs per Product Line—This value is the Manufacturing Cost per Unit multiplied by Units Produced from the Quantity Manufactured section. The Totals cell (G27) is the sum of cells D27, E27, and F27. You will use the value in the Totals cell in the Income Statement section.
- Sales Revenues per Product Line—This value is the Sales Price per bottle from the Constants section multiplied by Units Produced from the Quantity Manufactured section. The Totals cell

(G28) is the sum of cells D28, E28, and F28. You will use the value in this cell in the Income Statement section.

Income Statement Section

The last section you need to construct is the Income Statement, as shown in Figure D-48. An explanation of the needed formulas follows the figure.

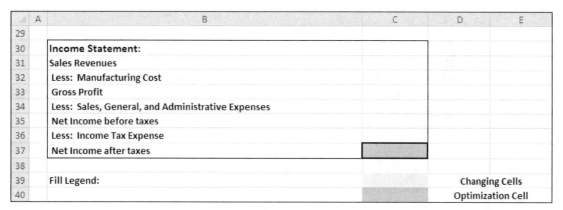

FIGURE D-48   Income Statement section with fill legend

- Sales Revenues—This value is the total sales revenues from the Calculations section (cell G28).
- Less: Manufacturing Cost—This value is the total manufacturing costs from the Calculations section (cell G27).
- Gross Profit—This value is the Sales Revenues minus the Manufacturing Cost.
- Less: Sales, General, and Administrative Expenses—This value is the Sales Revenues multiplied by the Sales, General, and Administrative Expenses per Dollar Revenue from the Constants section (cell C8).
- Net Income before taxes—This value is the Gross Profit minus the Sales, General, and Administrative Expenses.
- Less: Income Tax Expense—If the Net Income before taxes is greater than zero, this value is the Net Income before taxes multiplied by the Income Tax Rate in the Constants section. If Net Income before taxes is zero or less, the Income Tax Expense is zero.
- Net Income after taxes—This value is the Net Income before taxes minus the Income Tax Expense. You will use this value as your optimization cell because you want to maximize Net Income after taxes.

Setting up Solver

You need to satisfy the following conditions when running Solver:

- Your objective is to maximize Net Income after taxes (cell C37).
- Your changing cells are the Units Produced (cells D22, E22, and F22).
- Observe the following constraints:

  - You must produce at least the Minimum Sales Demand for each product (cells D6, E6, and F6).
  - Your total Lbs. of Ambergris Used (cell G25) cannot exceed the Available Ambergris (cell C9).
  - You cannot produce negative units of any product (enter constraints for the changing cells to be greater than or equal to zero).
  - You can produce only whole units of any product (enter constraints for the changing cells to be integers).

Run Solver and create an Answer Report when Solver finds the solution. When you complete the program, print your spreadsheet with the Solver solution, and print the Answer Report. Save your work and close Excel.

TROUBLESHOOTING SOLVER

Solver is a fairly complex software program. This section helps you address common problems you may encounter when attempting to run Solver.

Using Whole Numbers in Changing Cells

Before you run your first Solver model or rerun a previous model, always enter a positive whole number in each of the changing cells. If you have not already defined maximum and minimum constraints for the values in the changing cells, enter 1 in each cell before running Solver.

Getting Negative or Fractional Answers

If you receive negative or fractional answers when running Solver, you may have neglected to specify one or more of the changing cells as non-negative integers. Alternatively, if you are working on a cost minimization problem and you fail to specify the optimization cell as non-negative, you may receive a negative answer for the cost. Sometimes Solver will also warn you that you have one or more unbounded constraints (see Figure D-49).

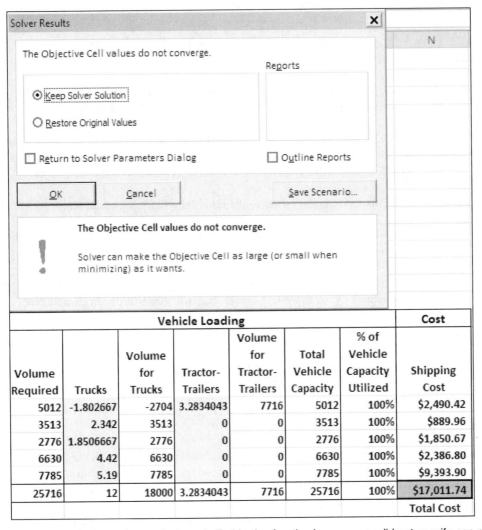

FIGURE D-49   Solver has an "unbounded" objective function because you did not specify non-negative integer constraints

Creating Overconstrained Models

If Solver cannot find a solution because it cannot meet the constraints you defined, you will receive an error message. When this happens, Solver may even violate the integer constraints you defined in an attempt to find an answer, as shown in Figure D-50.

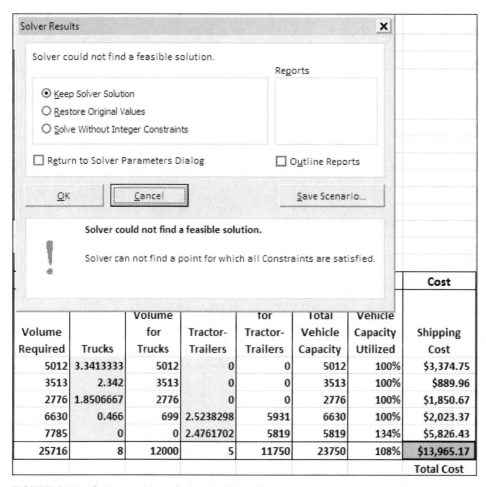

Volume Required	Trucks	Volume for Trucks	Tractor-Trailers	for Tractor-Trailers	Total Vehicle Capacity	Vehicle Capacity Utilized	Cost Shipping Cost
5012	3.3413333	5012	0	0	5012	100%	$3,374.75
3513	2.342	3513	0	0	3513	100%	$889.96
2776	1.8506667	2776	0	0	2776	100%	$1,850.67
6630	0.466	699	2.5238298	5931	6630	100%	$2,023.37
7785	0	0	2.4761702	5819	5819	134%	$5,826.43
25716	8	12000	5	11750	23750	108%	$13,965.17
							Total Cost

FIGURE D-50 Solver could not find a feasible solution because not enough vehicles were available

Setting a Constraint to a Single Amount

Sometimes you may want to enter an exact amount into a constraint, as opposed to a number in a range. For example, if you wanted to assign exactly 11 trucks in the CV Fitness problem instead of a maximum of 12, you would select the equals (=) operator in the Change Constraint window, as shown in Figure D-51.

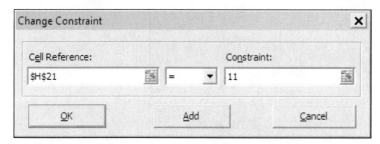

FIGURE D-51 Constraining a value to a specific amount

Setting Changing Cells to Integers

Throughout the tutorial, you were directed to set the changing cells to integers in the Solver constraints. In many business situations, there is a logical reason for demanding integer solutions, but this approach does have disadvantages. Forcing integers can sometimes increase the amount of time Solver needs to find a feasible solution. In addition, Solver sometimes can find a solution using real numbers in the changing cells instead of integers. If Solver cannot find a feasible solution or reports that it has reached its calculation time limit, consider removing the integer constraints from the changing cells and rerunning Solver to see if it finds an optimal solution that makes sense.

Restarting Solver with New Constraints

Suppose you want to start over with a completely new set of constraints. In the Solver Parameters window, click Reset All. You will be asked to confirm that you want to reset all the Solver options and cell selections (see Figure D-52).

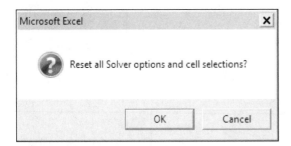

FIGURE D-52   Reset options warning

If you want to clear all the Solver settings, click OK. An empty Solver Parameters window appears with all the former entries deleted, as shown in Figure D-53. You can then set up a new model.

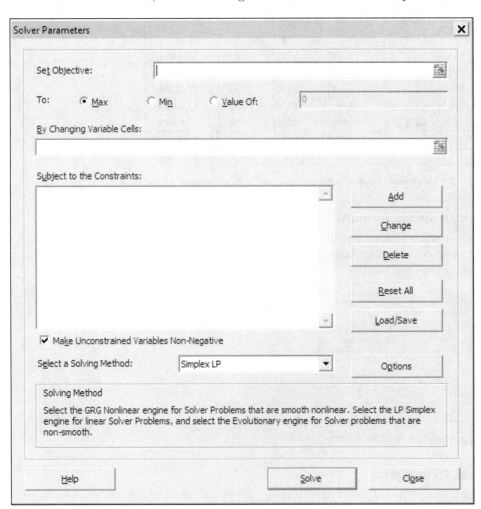

FIGURE D-53   Solver Parameters window after selecting Reset All

N O T E

Be certain that you want to select the Reset All option before you use it. If you only want to edit, delete, or add a constraint, use the Add, Delete, or Change button for that constraint.

Using the Solver Options Window

Solver has several internal settings that govern its search for an optimal answer. Click the Options button in the Solver Parameters window to see the default selections for these settings, as shown in Figure D-54.

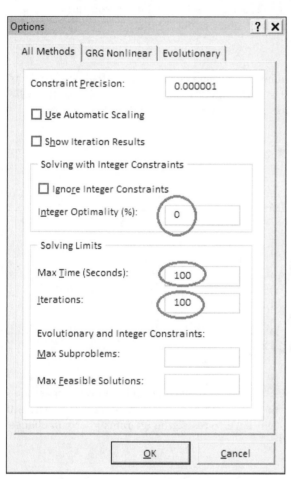

FIGURE D-54   Solver Options window

You should not need to change the settings in the Options window except for the default value of 5% for Integer Optimality. When it is set at 5%, Solver will get within 5% of the optimal answer, but this setting might not give you the lowest cost or highest income. Change the setting to 0 and click OK.

In more complex problems that have a dozen or more constraints, Solver may not find the optimal solution within the default 100 seconds or 100 iterations. If so, a window will prompt you to continue or stop (see Figure D-55). If you have time, click Continue and let Solver keep working toward the best possible solution. If Solver works for several minutes and still does not find the optimal solution, you can stop by pressing the Ctrl and Break keys together. Click Stop in the resulting window.

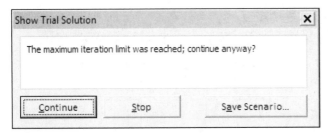

FIGURE D-55   Prompt that appears when Solver reaches its maximum iteration limit

If you think that Solver needs more time and iterations to reach an optimal solution, you can increase the Max Time and Iterations, but you should probably keep both values under 32,000.

Printing Cell Formulas in Excel

Earlier in the tutorial, you learned how to display cell formulas in your spreadsheet cells. Hold down the Ctrl key and then press the ~ key (on most keyboards, this key is next to the "1" key). You can change the cell widths to see the entire formula by clicking and dragging the column by the dividing lines between the column letters. See Figure D-56.

	R27		ƒ	Width: 14.14 (208 pixels)			
	G	H	I	J	K	L
13						
14				Vehicle Loading		
15	Volume Required	Trucks	Volume for Trucks	Tractor-Trailers	Volume for Tractor-Trailers	Total Vehicle Capacity
16	=D16*$C$8+E16*$C$9+F16*$C$10	2	=H16*$C$5	1	=J16*$C$6	=I16+K16
17	=D17*$C$8+E17*$C$9+F17*$C$10	1	=H17*$C$5	1	=J17*$C$6	=I17+K17
18	=D18*$C$8+E18*$C$9+F18*$C$10	2	=H18*$C$5	0	=J18*$C$6	=I18+K18
19	=D19*$C$8+E19*$C$9+F19*$C$10	3	=H19*$C$5	1	=J19*$C$6	=I19+K19
20	=D20*$C$8+E20*$C$9+F20*$C$10	4	=H20*$C$5	1	=J20*$C$6	=I20+K20
21	=SUM(G16:G20)	=SUM(H16:H20)	=SUM(I16:I20)	=SUM(J16:J20)	=SUM(K16:K20)	=SUM(L16:L20)

FIGURE D-56   Spreadsheet with formulas displayed in the cells

To print the formulas, click the File tab and select Print. To restore the screen to its normal appearance and display values instead of formulas, press Ctrl+~ again; the key combination is actually a toggle switch. If you changed the column widths in the formula view, you might have to resize the columns after you change back.

"Fatal" Errors in Solver

When you run Solver, you might sometimes receive a message like the one shown in Figure D-57.

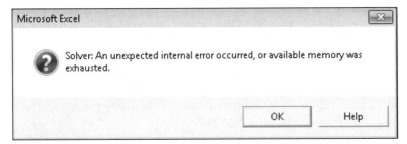

FIGURE D-57   Fatal error in Solver

Solver usually attempts to find a solution or reports why it cannot. When Solver reports a fatal error, the root cause is difficult to troubleshoot. Possible causes include merged cells on the spreadsheet or printing multiple Answer Reports after running Solver multiple times. A common solution to this error has been to remove the Solver add-in, close Excel, reopen it, and then reinstall Solver. If you encounter a fatal error when using this book, check with your instructor.

Sometimes Solver will generate strange results. Even when your cell formulas and constraints match the ones your instructor has created, Solver's answers might not match the "book" answers. You might have entered your constraints into Solver in a different order, you may have changed some of the options in Solver, or you may have specified real numbers instead of integers for the constraints (or vice versa). Also, the solving method you selected and the amount of time you gave Solver to work can affect the final answer. If your solution is close to the one posted by your instructor, but not exactly the same, show the instructor your setup in the Solver Parameters window. Solver is a powerful tool, but it is not infallible—ask your instructor for guidance if necessary.

DESERT AIRLINES AIRCRAFT ASSIGNMENT PROBLEM

Decision Support Using Microsoft Excel Solver

PREVIEW

Desert Airlines is a regional passenger airline and air freight service based in Las Vegas. The airline offers commercial flights to Palm Springs, Tucson, Phoenix, San Diego, Reno, and Los Angeles. Desert Airlines has grown in the past decade, and recently expanded its aircraft fleet to accommodate increased passenger demand. Herb Kellerman and June Rawlinson, the founders of Desert Airlines, are concerned that their manual method of daily aircraft assignment does not provide the best solution for operations. Tanya Rogers, the controller, thinks that the demands of the air freight service might be eating into the company's profitability. She has suggested the possibility of "tolling" some of the airline's air freight business to a separate air parcel company instead. *Tolling* is paying a competitor to provide a product or service to your company.

You have been hired as an MIS consultant to develop a DSS model for Desert Airlines. Your completed model will be used to assign the airline's fleet to its six destinations at a minimal cost. Furthermore, you will modify the model to examine how tolling some of the airline's freight service will affect operating costs and profitability.

PREPARATION

- Review the spreadsheet concepts discussed in class and in your textbook.
- Your instructor may assign Microsoft Excel exercises to help prepare you for this case.
- Tutorial D explains how to set up and use Solver for maximization and minimization problems. The CV Fitness exercise in the tutorial should be particularly helpful.
- Review the file-saving instructions in Tutorial C—saving an extra copy of your work on a USB thumb drive is always a good idea.
- Review Tutorial F to brush up on your presentation skills.

BACKGROUND

You will use your Excel skills to build a decision model and determine how many of each type of aircraft should be assigned to the six Desert Airlines destinations. The model requires the following data, which the management team has compiled:

- Data for the three types of aircraft in the Desert Airlines fleet, including the following:
  - Passenger capacity
  - Air freight capacity
  - Operating cost per mile (includes fuel, labor, and allocated overhead)
  - Number of available aircraft
- Ticket pricing to each destination
- Air freight pricing to each destination
- Air distance from Las Vegas to each destination

In addition, the marketing department has given you information about the passenger and air freight demand for each destination city:

- The expected daily passenger bookings
- The expected daily air freight demand in pounds

To satisfy passenger requirements and air freight demand, your Solver model will assign aircraft by number and type to each destination city. The model will also calculate total daily revenues from both passenger service and air freight, as well as the total daily operating cost. The results of these calculations will be used to create a daily gross profit statement. You will run Solver first to minimize the total operating cost. Next, you will modify the model to examine tolling some of the air freight business and its effect on the total operating cost. Finally, you will run the modified model to maximize daily gross profits.

Desert Airlines Aircraft Fleet

Desert Airlines started with a fleet of 12 Embraer 170 short-range passenger jets. Over the airline's decade of growth, 24 more Embraer jets were added to the fleet, but rapidly growing passenger demand required the addition of larger, more efficient aircraft. So, Desert Airlines purchased 32 Boeing 737-100 jets and 18 Airbus A320s. Almost all the planes are used daily, putting a strain on maintenance requirements and fleet availability. Herb and June would like to retire some of the older, less cost-efficient Embraers without jeopardizing customer service.

ASSIGNMENT 1: CREATING SPREADSHEET MODELS FOR DECISION SUPPORT

In this assignment, you will create spreadsheets that model the business decision Desert Airlines is seeking. In Assignment 1A, you will create a spreadsheet and attempt to assign the aircraft manually to minimize the total operating cost. In Assignment 1B, you will copy the spreadsheet to a new worksheet, and then set up and run Solver to minimize the total operating cost. In Assignment 1C, you will copy the Solver solution to a new worksheet and modify it to add calculations for tolling excess air freight. You will then rerun Solver to determine if tolling can further improve the airline's total operating cost. In Assignment 1D, you will copy the modified Solver spreadsheet and rerun Solver to maximize daily gross profit.

This section helps you set up each of the following spreadsheet components before entering the cell formulas:

- Constants
- Calculations and Results
- Income Statement

The Calculations and Results section is the heart of the decision model. You will set up columns for air distance, daily demand, aircraft assignment by type, aircraft use, and operating costs. The spreadsheet rows will represent destination cities. The Aircraft Assignment section will be the range of changing cells for Solver to manipulate. The total operating cost in this section will serve as your optimization cell for Assignments 1B and 1C, and the daily gross profit will be your optimization cell for Assignment 1D. You will add formulas to the air freight tolling cells in the Calculations and Results section for Assignments 1C and 1D.

Assignment 1A: Creating the Spreadsheet—Base Case

A discussion of each spreadsheet section follows. This information helps you set up each section of the model and learn the logic of the formulas in the spreadsheet. If you choose to enter the data directly, follow the cell structure shown in the figures. *You can also download the spreadsheet skeleton if you prefer.* To access the base spreadsheet skeleton, go to your data files, select Case 8, and then select **Desert Airlines Skeleton.xlsx**.

Constants Section

First, build the skeleton of your spreadsheet. Set up the spreadsheet title and Constants section as shown in Figure 8-1. An explanation of the column items follows the figure.

FIGURE 8-1 Spreadsheet title and Constants section

- Spreadsheet title—Enter the spreadsheet title in cell B1 and then merge and center the title across cells B1 through F1.
- Constants section, Aircraft Data table—Enter the column headings shown in cells B5 through G5.
- Aircraft Type—Enter each of the three aircraft listed in cells B6 through B8.
- Passenger Capacity—Enter each of the three passenger capacities listed in cells C6 through C8.
- Air Freight Cargo Capacity (lbs)—Enter each of the three cargo capacities listed in cells D6 through D8.
- Operating Cost per mile—Enter each of the three operating costs per mile listed in cells E6 through E8.
- Operating Cost per Passenger-Mile—This value is the Operating Cost per mile divided by the passenger capacity of the plane, but you do not need to enter formulas. Enter each of the three operating costs per passenger mile listed in cells F6 through F8. These values are not used in the Solver solution, but are provided as an aid to completing Assignment 1A.
- Available Fleet—This value is the number of planes of each type that Desert Airlines keeps in service. Enter these numbers in cells G6 through G8.
- Constants section, Fee Schedule table—Enter the column headings shown in cells B10 through D10.
- Destination—Enter the six destination cities in cells B11 through B16.
- Average Ticket Price—Enter the passenger ticket prices for the six destinations in cells C11 through C16.
- Air Freight Price/lb.—Enter the air freight price per pound for the six destinations in cells D11 through D16.
- Fill Legend—This section is actually adjacent to the Constants section. Enter "Fill Legend" in cell I5, fill cell I6 in yellow, fill cell I7 in orange, enter "Changing Cells" in cell J6, and enter "Optimization Cell" in cell J7.

Calculations and Results Section

The Calculations and Results section (see Figure 8-2) will contain air mileages, daily passenger bookings, and daily air freight shipment data obtained from the marketing department. Although these values are constants, keeping them in the Calculations and Results section facilitates writing and copying formulas in the Aircraft Utilization, Costs, and Tolling columns. This section also includes the Aircraft Assignment table, which contains the changing cells, and calculations for aircraft usage, costs, and tolling. An explanation of the sections and columns follows the figure.

	Calculations and Results Section:		Daily Demand		Aircraft Assignment			Aircraft Utilization				Costs	Tolling
	Destination	Distance from Las Vegas Hub	Daily Passenger Bookings	Daily Air Freight Shipments Lbs	Airbuses Assigned	Boeings Assigned	Embraers Assigned	Total Passenger Capacity	% of Passenger Capacity Utilized	Total Air Freight Capacity	% of Air Freight Capacity Utilized	Operating Cost	Lbs Air Freight to be Tolled
20	Palm Springs	193	1250	25000	1	1	1						
21	Reno	326	400	19000	1	1	1						
22	Phoenix	260	1840	78000	1	1	1						
23	Tucson	377	410	21450	1	1	1						
24	Los Angeles	226	2350	180000	1	1	1						
25	San Diego	278	1150	67000	1	1	1						
26		Total/Avg											
27												Total Cost	

FIGURE 8-2 Calculations and Results section

- Table headings—If you did not use the skeleton spreadsheet, enter the column headings shown in cells B18 through N19 in Figure 8-2.
- Destination—Cells B20 through B25 hold the six cities serviced daily by Desert Airlines.
- Distance from Las Vegas Hub—Cells C20 through C25 hold the distances in air miles to each of the six destinations.
- Daily Passenger Bookings—Cells D20 through D25 hold the average number of passenger tickets booked each day.
- Daily Air Freight Shipments (lbs.)—Cells E20 through E25 hold the average number of pounds of freight shipped daily.
- Aircraft Assignment section—Cells F20 through H25 are the heart of the Solver model—the changing cells. The cells hold the amounts of each of the three aircraft types that Solver will assign to the six destinations. Enter "1" in each of these cells for now. You should fill the cells with a background color to indicate that they are the changing cells for Solver. To fill the cells, select them and then click the Fill Color button in the Font group on the Home tab. (The button's icon is a can of paint.) In the spreadsheet skeleton, the cells are yellow.
- Aircraft Utilization section, Total Passenger Capacity—Cells I20 through I25 hold the total passenger capacity for each destination. The capacity is calculated by multiplying the number of each assigned aircraft type by its passenger capacity, which is taken from cells C6 through C8 of the Constants section. Next, take the sum of the total capacities for the three types of aircraft assigned. Be sure to use *absolute* cell references for the passenger capacity values from the Constants section so that you have to write the formula only for the first cell (I20); then you can copy the formula to cells I21 through I25.
- Percent of Passenger Capacity Utilized—Cells J20 through J25 hold the percentage of passenger capacity used for each destination. The value is calculated by dividing Daily Passenger Bookings by Total Passenger Capacity.
- Total Air Freight Capacity—Cells K20 through K25 hold the total air freight capacity for each destination. The capacity is calculated by multiplying the number of each assigned aircraft type by its freight cargo capacity, which is taken from cells D6 through D8 of the Constants section. Next, you take the sum of the total capacities for the three types of aircraft assigned. Again, you must use *absolute* cell references for the freight capacity values from the Constants section so that you only have to write the formula for the first cell (K20); then you can copy the formula to the other five cells.
- Percent of Air Freight Capacity Utilized—Cells L20 through L25 hold the percentage of air freight capacity used. The percentage is calculated by dividing the Daily Air Freight Shipments by the Total Air Freight Capacity.
- Operating Cost—Cells M20 through M25 hold the operating cost for each aircraft type to each destination. The cost is calculated by the following formula:

Number of aircraft assigned × Operating Cost per mile × Mileage to destination

The costs for each of the three aircraft are then added to get the operating cost. Remember to use *absolute* cell references for the Operating Cost per mile, which is listed in cells E6 through E8 of the Constants section. That way, you have to write the formula only for the first cell (M20); then you can copy the formula to the other five cells.

- Lbs. Air Freight to be Tolled—Leave cells N20 through N25 blank for now. You will place formulas in these cells in Assignment 1C.

- Total/Avg—Total the cell entries for every column *except* column J (% of Passenger Capacity Utilized) and column L (% of Air Freight Capacity Utilized). Place the totals in cells D26 to I26, K26, and M26. For cell J26, divide the total Daily Passenger Bookings in cell D26 by the Total Passenger Capacity in cell I26 and format the result as a percentage. For cell L26, divide the total Daily Air Freight Shipments in cell E26 by the Total Air Freight Capacity in cell K26 and format the result as a percentage. These two averages are the overall utilization rates of your fleet. For now, leave cell N26 blank (total for Lbs. Air Freight to be Tolled).

If you wrote your formulas correctly, the Calculations and Results section should look like Figure 8-3.

	A	B	C	D	E	F	G	H	I	J	K	L	M	N
17														
18		Calculations and Results Section:		Daily Demand		Aircraft Assignment			Aircraft Utilization				Costs	Tolling
19		Destination	Distance from Las Vegas Hub	Daily Passenger Bookings	Daily Air Freight Shipments Lbs	Airbuses Assigned	Boeings Assigned	Embraers Assigned	Total Passenger Capacity	% of Passenger Capacity Utilized	Total Air Freight Capacity	% of Air Freight Capacity Utilized	Operating Cost	Lbs Air Freight to be Tolled
20		Palm Springs	193	1250	25000	1	1	1	340	368%	18000	139%	$18,142	
21		Reno	326	400	19000	1	1	1	340	118%	18000	106%	$30,644	
22		Phoenix	260	1840	78000	1	1	1	340	541%	18000	433%	$24,440	
23		Tucson	377	410	21450	1	1	1	340	121%	18000	119%	$35,438	
24		Los Angeles	226	2350	180000	1	1	1	340	691%	18000	1000%	$21,244	
25		San Diego	278	1150	67000	1	1	1	340	338%	18000	372%	$26,132	
26			Total/Avg:	7400	390450	6	6	6	2040	363%	108000	362%	$156,040	
27													Total Cost	

FIGURE 8-3 Completed Calculations and Results section

Income Statement Section

The Income Statement section (see Figure 8-4) is actually a projection of daily gross profits, and is based on the number of aircraft that will be assigned either manually or by Solver. An explanation of the line items follows the figure.

	A	B	C
27			
28		Income Statement Section:	
29		Passenger Revenues:	
30		Air Freight Revenues:	
31		Total Revenues:	
32		less Operating Costs:	
33		less Tolling Cost:	
34		Daily Gross Profit:	

FIGURE 8-4 Income Statement section

- Passenger Revenues—This value is calculated by multiplying the passenger tickets booked for each destination (cells D20 through D25) by their respective average ticket prices (cells C11 through C16), and then totaling the ticket revenues for the six destinations.
- Air Freight Revenues—This value is calculated by multiplying the daily air freight shipments for each destination (cells E20 through E25) by their respective air freight prices (cells D11 through D16), and then totaling the air freight revenues for the six destinations.
- Total Revenues—This value is the total of Passenger Revenues and Air Freight Revenues.
- Less Operating Costs—This value is the Total Cost from cell M26.
- Less Tolling Cost—This value is the Tolling cost from cell N26, which will be used later in the assignment.
- Daily Gross Profit—This value is the Total Revenues minus the Operating Costs and the Tolling Cost. This cell will be used as the optimization cell for Assignment 1D.

If your formulas are correct, the initial Income Statement section will appear as shown in Figure 8-5.

	A	B	C
27			
28		Income Statement Section:	
29		Passenger Revenues:	$1,049,810
30		Air Freight Revenues:	$197,383
31		Total Revenues:	$1,247,193
32		less Operating Costs:	$156,040
33		less Tolling Cost:	
34		Daily Gross Profit:	$1,091,153

FIGURE 8-5   Initial Income Statement

The initial income statement correctly reflects the revenues expected from the passenger and air freight bookings, but the operating costs are not correct because the aircraft required to transport the passengers and freight have not been completely assigned yet.

Attempting a Manual Solution

Attempt to assign your aircraft fleet manually in the spreadsheet. You have several good reasons for doing this. First, you can make sure your model is working correctly before you set up Solver to run. Second, assigning the aircraft fleet manually will demonstrate which constraints you must meet in solving the problem. For instance, if a passenger or air freight utilization rate is over 100%, you have not assigned enough planes to carry all the passengers or air freight to a particular destination. Therefore, one constraint is that the total passenger capacity for the planes assigned to a destination must be greater than or equal to the passenger bookings. Another constraint is that the total air freight capacity for the planes assigned must be greater than the air freight shipments booked. Given the fleet size, you can probably assign the fleet manually and meet all of your constraints. However, will your total operating cost be the least expensive solution? Running the problem manually will provide an initial operating cost to which you can compare your Solver solution later. The Solver optimization tool should give you a better solution than assigning the fleet manually.

When attempting to assign the aircraft manually in the Aircraft Assignment section (the changing cells), you want to assign the most cost-efficient planes first, which is why the Operating Cost per Passenger-Mile data is included in the Constants section. You will want to assign all the Airbus 300 planes first, then the Boeing 737s, and finally the Embraer 170s. You must satisfy *both* the passenger and air freight demands for each destination—in other words, the Total Passenger Capacity values in cells I20 through I25 and the Total Air Freight Capacity values in cells K20 through K25 must be equal to or greater than the Daily Demand values in cells D20 through D25 and E20 through E25. If you have satisfied the passenger and air freight demands correctly, none of the utilization rates in cells J20 through J25 and L20 through L25 will exceed 100%. In addition, the total aircraft assigned for each type (cells F26 through H26) cannot exceed the available number of each aircraft type (cells G6 to G8).

Once you reach a solution that satisfies the preceding constraints, save your workbook. Name the worksheet **Desert Airline Guess** and then right-click the worksheet name tab (see Figure 8-6). Click Move or Copy, then copy the worksheet (see Figure 8-7) and rename the new copy **Desert Airlines Solver 1**. You will use the new worksheet to complete the next part of the assignment.

FIGURE 8-6   Right-clicking the worksheet name tab

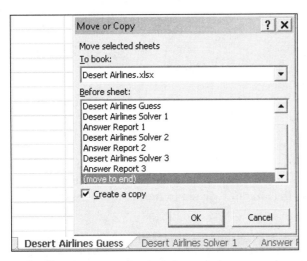

FIGURE 8-7 Move or Copy menu

Assignment 1B: Setting up and Running Solver

Before using the Solver Parameters window, you should jot down the parameters you must define and their cell addresses. Here is a suggested list:

- The cell you want to minimize (Total Cost, cell M26)
- The cells you want Solver to manipulate to obtain the optimal solution (Aircraft Assignment, cells F20 through H25)
- The constraints you must define:
  - All the aircraft assignment cells must be non-negative integers.
  - The total number of each type of aircraft assigned (cells F26 through H26) cannot exceed the number of available aircraft of each type (cells G6 through G8).
  - The total passenger capacity to each destination (cells I20 through I25) must be equal to or greater than the total passenger bookings for each destination (cells D20 through D25).
  - The total air freight capacity to each destination (cells K20 through K25) must be equal to or greater than the total air freight shipment for each destination (cells E20 through E25).

Next, set up your problem. In the Analysis group on the Data tab, click Solver; the Solver Parameters window appears, as shown in Figure 8-8. Enter "Total Cost" in the Set Objective text box, click the Min button, designate your Changing Cells (cells F20 through H25), and add the constraints from the preceding list. Use the default Simplex LP solving method. If you need help defining your constraints, refer to Tutorial D.

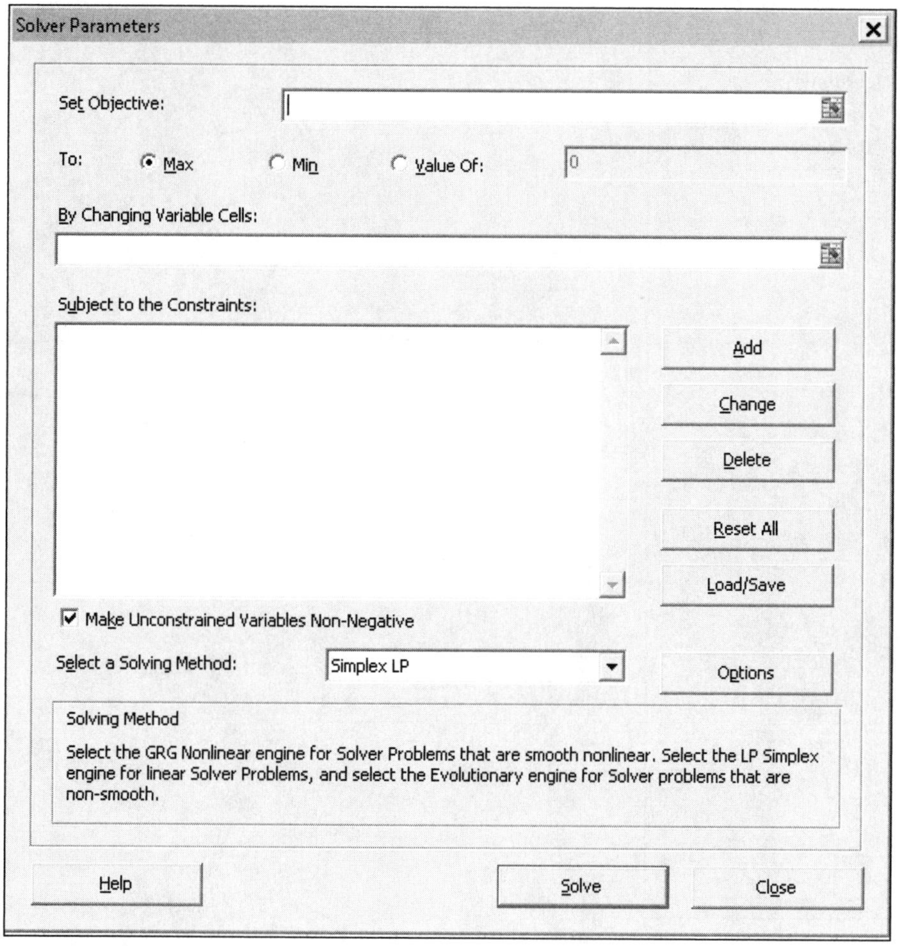

FIGURE 8-8 The Solver Parameters window

Next, you should click the Options button and check the Options window that appears (see Figure 8-9). The default Integer Optimality is 5%; change it to 1% to get a better answer. Make sure the Constraint Precision is set to the default value of .000001 and that the Use Automatic Scaling option is checked. When you finish setting the options, click OK to return to the Solver Parameters window.

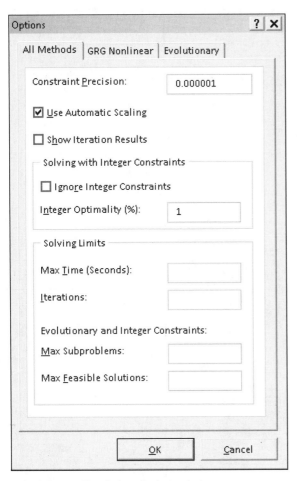

FIGURE 8-9 The Solver Options window

Run Solver and click Answer Report when Solver finds a solution that satisfies the constraints. When you finish, print the entire workbook, including the Solver Answer Report Sheet. To save the workbook, click the File tab and then click Save. For the rest of the case, you either can use the Save As command to create new Excel workbooks or continue copying and renaming the worksheets. Both options offer distinct advantages, but having all of your worksheets and Solver Answer Reports in one Excel workbook allows you to compare different solutions easily, as well as prepare summary reports.

Before continuing, examine the aircraft assignments that Solver chose for minimizing the total cost. If you set up Solver correctly, you should see a significant reduction in total cost from your manual assignment. You should also see that Solver assigned fewer aircraft, especially from the older Embraer 170 fleet. Therefore, Desert Airlines might be able to remove some of the older aircraft from service or resell them.

Assignment 1C: Tolling the Excess Air Freight with Solver

Using the model you created in Assignment 1A, Desert Airlines can also determine the cost benefit of having an air freight service "toll" excess air freight. As you learned earlier, tolling means paying a competitor to provide a product or service to your company. Tolling might be a good option for Desert Airlines if its air freight service requires more aircraft than necessary to meet the passenger bookings. Recall that company controller Tanya Rogers suspects that the less profitable air freight service might be hurting the airline's profitability. After contacting large air freight companies that service the same six destinations from Las Vegas, the best quote she obtained for air freight was a flat rate of $1 per pound. You must modify the worksheet to accommodate tolling the excess air freight and then rerun Solver.

Copy your Desert Airlines Solver 1 worksheet and rename it **Desert Airlines Solver 2**. Before running Solver again, you must perform the following tasks:

1. You must insert a calculation for the cells in the Tolling column (cells N20 through N26).
2. You must link the total air freight tolled in cell N26 to the Tolling Cost cell in the Income Statement section (cell C33).
3. You must remove some of the constraints in the Solver setup.

Note that when you copy a worksheet, the Solver Parameters window sometimes might attempt to refer back to the old one. When this happens, the following error message appears: "Objective Cell must be a single cell on the active sheet." When you click OK to close the error message, the Solver Parameters window opens, but the values look like the following: "='Desert Airlines Guess'!#REF!." The message means that Solver cannot find the correct cell in the current worksheet. Unfortunately, the only solution is to click Reset All in the Solver Parameters window and reenter the Solver parameters.

Using tolling, a company still gets revenues from business with its customers, but at the same time it pays a competitor for providing the business services. Tolling allows Desert Airlines to assign aircraft to meet only the required passenger bookings. If the aircraft assigned to carry passengers do not have the capacity to carry all of the air freight shipments, then the excess over total aircraft capacity will be tolled to the competitor. The amount of air freight that must be tolled to each destination depends on the aircraft assignments made by Solver.

You must use an IF statement to write the formulas for cells N20 through N26. Use the following logic for the formula you place in cell N20:

- If the Total Air Freight Capacity to the Palm Springs destination (cell K20) is equal to or greater than the Daily Air Freight Shipments (cell E20), then no air freight needs to be tolled (cell N20 equals zero).
- If the Total Air Freight Capacity to the Palm Springs destination (cell K20) is less than the Daily Air Freight Shipments (cell E20), then the amount to be tolled (cell N20) equals the Daily Air Freight Shipments minus the Total Air Freight Capacity.

Click the Insert Function button (f_x) next to the formula bar below the Ribbon to help you write the IF statement. See Figure 8-10. After you enter the formula in cell N20, you can copy and paste the formula into cells N21 through N25. Cell N26 is the total of cells N20 through N25.

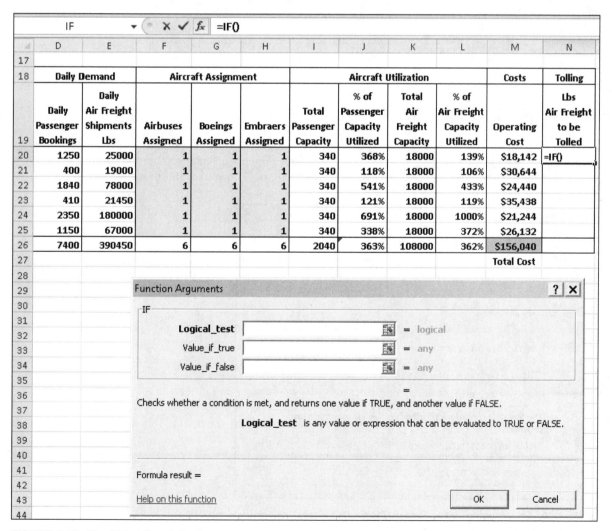

FIGURE 8-10 The IF function dialog box

Next, enter "=N26" in cell C33 of the Income Statement section. Because the tolling cost is $1 per pound, you do not need to enter a formula to multiply the pounds tolled by the tolling cost per pound. If the cost were less or more than $1, you would have to include the cost in the Constants section and then multiply the pounds tolled by the cost per pound to get the total tolling cost.

Finally, you must redefine the constraints in the Solver model. When you copy a worksheet, its defined Solver parameters are copied along with it. Therefore, when you return to the Solver Parameters window, it should contain all the values from the first Solver run. Click each of the six constraints for the air freight, then click the Delete button for each highlighted constraint to remove it. Run Solver again—when it finds the solution, click Answer Report and accept the solution. Solver creates a report in a new worksheet and names it Answer Report 2.

Examine the new Solver solution, comparing both the total operating costs and gross profit to the earlier Solver solution. Although you now have to pay tolling costs, did the operating cost savings offset the tolling cost? In other words, is gross profit higher than in the earlier solutions? If the gross profit is higher, then tolling the excess air freight is a good business decision. However, this raises another question: Should you run Solver again to maximize gross profit instead of minimizing operating costs? Because you added tolling to the business model, you should probably maximize gross profit to account for the effect of tolling.

Assignment 1D: Rerunning Solver to Maximize Gross Profit

Copy the worksheet that contains the Solver solution for tolling, and rename the new worksheet **Desert Airlines Solver 3**. Click Solver to open the Solver Parameters window, and then change the value in the Set Objective text box to $C$34 (the cell that contains Daily Gross Profit). Click the Max button to maximize the Set Objective value, as shown in Figure 8-11, and then run Solver.

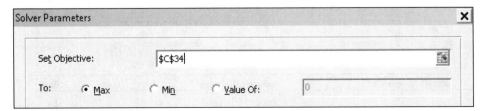

FIGURE 8-11   Changing the objective and solving for maximized profit

The Solver Results window appears (see Figure 8-12) and warns that the linearity conditions required by the Simplex LP solving method are not satisfied. The default Simplex calculation method will not work because the model now includes the tolling equations, which are IF formulas and therefore nonlinear.

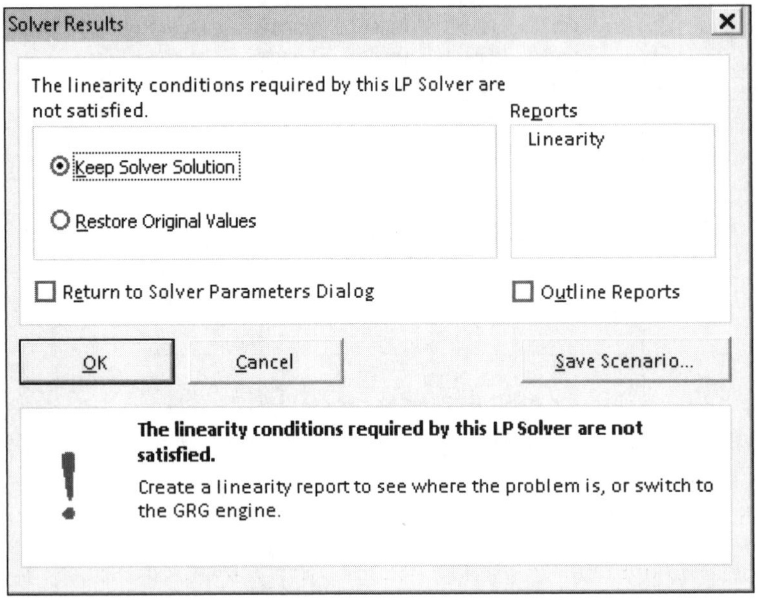

FIGURE 8-12   Error message when Simplex LP method is used on nonlinear models

Fortunately, Solver has methods for working with nonlinear problems as well. Open the Solver Parameters window again and click the Select a Solving Method list arrow. Click GRG Nonlinear (see Figure 8-13), then click Options. Under Options, click the All Methods tab to check that the Integer Optimality is set to 1% (see Figure 8-14), then click OK to return to the Solver Parameters window. Run Solver again; the solution will probably take longer than in the earlier problems. Click Answer Report and then click OK to create a third Answer Report. Examine it and the worksheet to see if maximizing daily gross profit is a better solution than minimizing total cost.

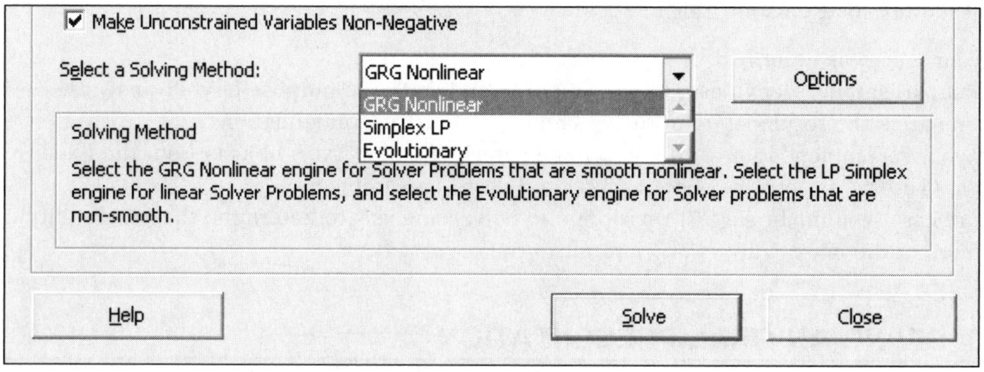

FIGURE 8-13   Selecting the GRG Nonlinear solving method

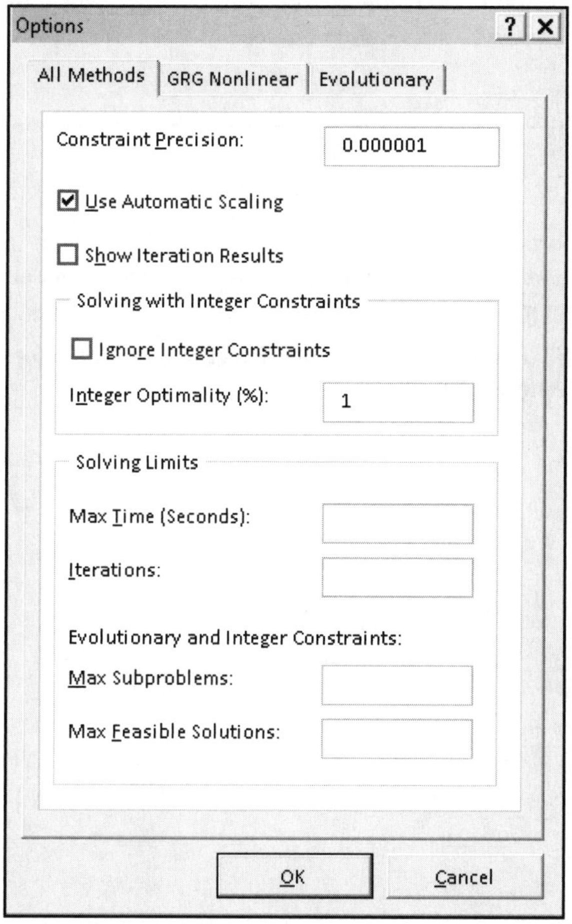

FIGURE 8-14   Solver Options window with Integer Optimality set to 1%

Print the Desert Airlines Solver 3 worksheet and Answer Report 3 and then save your workbook.

ASSIGNMENT 2: USING THE WORKBOOK FOR DECISION SUPPORT

You have built a series of worksheets to determine the best aircraft assignments with and without tolling excess air freight. You will now complete the case by using your solutions and Answer Reports to make recommendations in a memorandum.

Use Microsoft Word to write a memo to the management team at Desert Airlines. State the results of your analysis and state whether you think tolling excess air freight is a sound business decision. In addition, you should recommend how many of the older Embraer aircraft to retire from service, depending on whether Desert Airlines decides to toll its excess air freight.

- Set up your memo as described in Tutorial E.
- In the first paragraph, briefly describe the situation and state the purpose of your analysis.
- Next, summarize the results of your analysis and give your recommendations.
- Support your recommendation with appropriate screen shots or Excel objects from the Excel workbook. (Tutorial C describes how to copy and paste Excel objects.)
- In a future case, you might suggest revisiting the Solver analysis to determine the profitability of adding more of the newer Airbus 300 aircraft to the current fleet.

ASSIGNMENT 3: GIVING AN ORAL PRESENTATION

Your instructor may request that you summarize your analysis and recommendations in an oral presentation. If so, prepare a presentation for the CEO and other managers that lasts 10 minutes or less. When preparing your presentation, use PowerPoint slides or handouts that you think are appropriate. Tutorial F explains how to prepare and give an effective oral presentation.

DELIVERABLES

Prepare the following deliverables for your instructor:

1. A printout of the memo
2. Printouts of your worksheets and Answer Reports
3. Your Word document, Excel workbook, and PowerPoint presentation on electronic media or sent to your course site as directed by your instructor

Staple the printouts together with the memo on top. If you have more than one Excel workbook file for your case, write your instructor a note that describes the different files.

PREMIER PLASTICS LINE SCHEDULING PLAN

Decision Support Using Microsoft Excel Solver

PREVIEW

Premier Plastics is a privately owned manufacturer of flexible packaging film, and has several plants across the country. The company's plant in western Maryland produces rolls of plastic film that are used to package bakery products, paper towels, and bathroom tissue.

To make plastic film, plastic pellets are fed into extruders, which melt the pellets into a viscous liquid. The molten plastic is then pumped through a slot die and quickly drawn into a web of thin film. The film is wound at high speed into large-diameter rolls called master rolls, which are slit into smaller rolls at customer-ordered widths for later use in packaging machines. Premier's Maryland plant has two plastic film lines: Line 41, which can produce a master roll that is 75 inches wide, and Line 42, which can produce a 105" wide master roll. (Inches are often indicated by double quotation marks.) Premier's customers order rolls of film in five different widths: 30", 35", 40", 45", and 50".

Fran Peters is the production scheduler at Premier Plastics. She schedules Lines 41 and 42 to meet customer demand, and she must determine the best way to cut the master rolls into widths ordered by customers without leaving too much scrap. She has scheduled manually for several years and has always kept the slitting yields above 90%; that is, less than 10% of the master rolls end up as scrap. However, Fran thinks that a computerized system would better optimize the slitting patterns to minimize waste and cost.

You are the corporate MIS manager for Premier Plastics. You have traveled to the Maryland plant to meet with Fran and develop an optimized scheduling program in Microsoft Excel to use for Lines 41 and 42.

PREPARATION

- Review the spreadsheet concepts discussed in class and in your textbook.
- Your instructor may assign Excel exercises to help prepare you for this case.
- Tutorial D covers how to set up and use Solver for maximization and minimization problems.
- Review the file-saving instructions in Tutorial C—saving an extra copy of your work on a USB thumb drive is always a good idea.
- Review Tutorial F to brush up on your presentation skills.

BACKGROUND

You will use your Excel skills to build a decision model that determines the best combination of slitting patterns to run on Lines 41 and 42 and meet customer demand for rolls of different widths. These customer demands in the spreadsheet skeleton represent the annual sales demand for the Maryland plant.

Fran has already determined the slitting patterns that provide the best yields for the two lines and has created a table for you to use in the Excel spreadsheet. The table shows how many finished rolls can be produced from a master roll of each given width (75" for Line 41 and 105" for Line 42). For example, a 75" master roll can produce one 40" finished roll and one 35" finished roll with zero waste. The same master roll can produce two 35" finished rolls with 5" of waste. Each inch of waste from the master roll weighs ten pounds, so five inches of waste would weigh 50 pounds. The waste film from the slitting pattern is reground and fed back into the extrusion process, but there is a limit to how much waste can be added to the process. As long as the finished yields are above 80%, Premier can reprocess all the slitting waste.

Fran has been a production scheduler at the Maryland plant for several years, and she realizes that the master roll widths of Lines 41 and 42 may not offer the best match to customer demands for finished roll widths. She has done preliminary work with the corporate engineering staff to obtain funds to purchase and install a new line, and to determine the possible slitting patterns for a 120" line. As a result, she thinks a state-of-the-art extrusion line that produces 120" master rolls might be more cost efficient than running the two older lines. After you build the Excel scheduling model for the existing lines, Fran wants you to modify the model to examine the economics of replacing the two current lines with the proposed new line.

The corporate Engineering and Accounting departments have given you information about the performance and manufacturing costs of Lines 41 and 42. The information includes the following:

- The cost per pound of the raw material
- The direct labor cost per machine hour for each line
- The manufacturing overhead cost allocated per machine hour for each line
- Each line's production rate in pounds per hour
- The master roll weight in pounds for each line
- The available machine hours per year for each line

The Marketing and Sales Department has provided the annual sales forecast by customer roll width. As you read previously, Fran has developed a table of slitting patterns for the five customer roll widths that can be run on Line 41 or 42. This table is included in the spreadsheet skeleton.

ASSIGNMENT 1: CREATING A SPREADSHEET FOR DECISION SUPPORT

In this assignment, you will create a spreadsheet that models Premier's line scheduling plan for Lines 41 and 42, with an emphasis on minimizing total scrap and total manufacturing cost. In Assignment 1A, you will create a spreadsheet to model the manufacturing plan and attempt to assign the slitting patterns manually. Then you will set up Solver to minimize the slitting scrap. In Assignment 1B, you will copy your model to a new worksheet and set up Solver to minimize total manufacturing cost. In Assignment 1C, you will set up a new Solver spreadsheet to examine the financial impact of a proposed 120" line to replace the two current lines. In Assignments 2 and 3, you will use data from the spreadsheet models to summarize the planned layouts and to provide a financial analysis and recommendation for the proposed new line. You might also be asked to prepare and give an oral presentation to support your analysis and recommendations.

The Premier Plastics spreadsheet model will contain the following sections:

- Constants
- Calculations and Results
- Cost Summary

In addition, the spreadsheet model for the proposed new line will contain a Capital Investment section.

Assignment 1A: Creating the Line Scheduling Spreadsheet

A discussion of each spreadsheet section follows. This information helps you set up each section of the model and learn the logic of the formulas. If you choose to enter the data directly, follow the cell structure shown in the figures. *To save time, it is highly recommended that you use the spreadsheet skeletons.* To access these spreadsheet skeletons, select Case 9 in your data files, and then select **Premier Plastics Skeletons.xlsx**. The file contains two skeleton worksheets. Copy the Premier Skeleton worksheet for Assignments 1A and 1B.

Constants Section

First, build the skeleton of your spreadsheet. Set up the spreadsheet title and Constants section as shown in Figure 9-1. An explanation of the line items follows the figure.

	A	B	C	D	E	F	G
1							
2		Premier Plastics--Packaging Film Production Scheduling					
3							
4		Constants Section:					
5			Line 41	Line 42			
6		Raw Material Cost per lb.	$0.42	$0.42			
7		Direct Labor Cost per Mach hour	$37	$58			
8		Mfg Overhead Cost per Mach hour	$450	$825			
9		Production Rate in lbs per hour	1500	3000			
10		Master Roll Weight in lbs	750	1050			
11		Available Machine Hours	2000	2000			

FIGURE 9-1 Spreadsheet title and Constants section

- Spreadsheet title—Enter the spreadsheet title in cell B1, then merge and center the title across cells B1 through H1.
- Raw Material Cost per lb.—This value is the cost per pound of plastic resin. The value is the same for both lines.
- Direct Labor Cost per Mach hour—This value is the direct labor cost to operate each line per machine hour.
- Mfg Overhead Cost per Mach hour—This value is the amount of manufacturing overhead cost allocated to each line per machine hour. Note that *Mfg* is a common abbreviation for *Manufacturing*; it is used throughout the case to shorten several titles and labels.
- Production Rate in lbs per hour—This value is the throughput in pounds per hour for each line. Although Line 42 is only 40% wider than Line 41, it has twice the throughput because it has bigger extruders and a faster winding capability.
- Master Roll Weight in lbs—This value is the weight of a master roll for each line. A master roll is a full-width roll that has been wound on the extrusion line but has not yet been slit down into finished rolls. The master rolls on both lines weigh 10 pounds per inch of width.
- Available Machine Hours—This value represents the total available hours per year for each extrusion line. Premier Plastics currently runs the machines 40 hours per week, with a two-week shutdown in the summer for maintenance.

Calculations and Results Section

This section is the heart of your decision model. It contains a table of the most efficient slitting patterns for the two lines based on the finished roll widths in the first five columns. The section also contains a column for the changing cells (the number of master rolls scheduled for each slitting pattern), and columns to calculate the waste for each slitting pattern scheduled.

Again, it is highly recommended that you use the skeleton file for this case. If you choose to set up the model yourself, refer to Figure 9-2. A description of each line item follows the figure. Note that cells filled with gray contain formulas, while cells filled with yellow are the changing cells for this model.

	B	C	D	E	F	G	H	I	J
13	Calculations and Results Section:								
15	Finished Roll Weight	500	450	400	350	300			
16	Finished Roll Widths	50"	45"	40"	35"	30"	No. of Master Rolls Scheduled	Scrap Wt per Master Roll	Lbs Scrap per Layout
17	Line 41 (75" wide)	0	1	0	0	1	1	0	
18	Possible Slitting Combinations	0	0	1	1	0	1	0	
19		1	0	0	0	0	1	250	
20		0	0	0	2	0	1	50	
21		0	0	1	0	1	1	50	
22		0	0	0	0	2	1	150	
23				Total Master Rolls Line 41					
24	Line 42 (105" wide)	2	0	0	0	0	1	50	
25	Possible Slitting Combinations	0	0	0	3	0	1	0	
26		0	0	0	2	1	1	50	
27		1	1	0	0	0	1	100	
28		1	0	1	0	0	1	150	
29		1	0	0	1	0	1	200	
30		0	0	2	0	0	1	250	
31		0	2	0	0	0	1	150	
32		0	0	1	0	2	1	50	
33		0	1	0	0	2	1	0	
34		0	0	0	0	3	1	150	
35				Total Master Rolls Line 42					
36								Total Scrap Weight lbs	
37	Finished Rolls by Width-Line 41								
38	Finished Rolls by Width-Line 42								
39	Total Finished Rolls by Width							RM Yield	
40	Sales Plan--Rolls by Width	1100	1230	3475	2800	5750			
42	Excess Production by Width								

FIGURE 9-2   Calculations and Results section

NOTE—SLITTING PATTERNS

Not all the possible slitting patterns for Lines 41 and 42 are listed in the model skeleton. Listing all the possible patterns would make the model unnecessarily complex, so only the most efficient patterns are displayed.

- Finished Roll Weight—Cells C15 through G15 hold the weight in pounds of one finished roll for each of the widths listed in the following set of cells.
- Finished Roll Widths—Cells C16 through G16 contain headings for the five roll widths that Premier Plastics sells.
- Possible Slitting Combinations—Cells C17 through G22 and C24 through G34 hold the number of finished rolls of each width that can be slit from one master roll (75" wide for Line 41 and 105" wide for Line 42). A total of 17 slitting combinations from both lines can result in scrap weights from zero to 250 pounds per master roll.
- No. of Master Rolls Scheduled—Cells H17 through H22 and H24 through H34 hold the number of master rolls scheduled for each of the possible slitting patterns for Lines 41 and 42. These cells are the heart of your model—they are the changing cells that Solver will manipulate, so they are filled in yellow on the skeleton spreadsheet. Cells H23 and H35 hold master roll totals for Lines 41 and 42, respectively. Place a "1" in each of these cells as a starting value so you can check the formulas when you finish the section.
- Scrap Wt per Master Roll—Cells I17 through I22 and I24 through I34 hold the number of pounds of scrap produced from one master roll slit to the pattern listed in that row. This value is the number of inches of scrap from the master roll multiplied by ten pounds per inch.
- Lbs Scrap per Layout—Cells J17 through J22 and J24 through J34 hold the scrap weight per master roll multiplied by the number of master rolls scheduled for each layout.
- Total Scrap Weight lbs.—Cell J36 holds the sum of cells J17 through J22 and cells J24 through J34. This value is the total of all scrap produced by the master rolls for each slitting pattern. This value also is the optimization cell that you will specify as the amount to minimize in your first Solver model, so the cell is filled in orange.
- RM Yield—Cell J38 holds the raw material (RM) yield. This value is calculated by subtracting the total scrap weight in cell J36 from the Total Lbs Raw Material Used in the Cost Summary section (cell E47), then dividing the result by the Total Lbs Raw Material Used. The answer should be displayed as a percentage.
- Finished Rolls by Width, Line 41—Cells C37 through G37 hold values for the number of finished rolls for Line 41. The formula for each cell is calculated by multiplying the number of finished rolls of a particular width by the number of master rolls scheduled for that pattern, and then summing the totals of all the patterns for Line 41. The formula is fairly long, so you can use the following formula in cell C37 to help you get started:

=C17*$H$17+C18*$H$18+C19*$H$19+C20*$H$20+C21*$H$21+C22*$H$22

- Finished Rolls by Width, Line 42—Cells C38 through G38 hold values for the number of finished rolls for Line 42. The formula for each cell is calculated by multiplying the number of finished rolls of a particular width by the number of master rolls scheduled for that pattern, and then summing the totals of all the patterns for Line 42. The formula is even longer than the previous one, so you can use the following formula in cell C38 to help you get started:

=C24*$H$24+C25*$H$25+C26*$H$26+C27*$H$27+C28*$H$28+C29*$H$29+ C30*$H$30 +C31*$H$31+C32*$H$32+C33*$H$33+C34*$H$34

- Total Finished Rolls by Width—Cells C39 through G39 hold the sum of the finished rolls by width for Line 41 and the finished rolls by width for Line 42.
- Sales Plan—Rolls by Width—Cells C40 through G40 hold the forecasted sales values in rolls for each width. The scheduled production must meet the sales plan; in other words, the total finished rolls of each width must be greater than or equal to the amounts in the sales plan.
- Excess Production by Width—Cells C42 through G42 represent finished rolls by width that are produced in excess of the sales plan. Because the rolls are in standard widths, they might be sold later and therefore should not be counted as scrap. If the total finished rolls by width are equal to the sales plan amounts for that width, then the excess production is zero. If the total finished rolls by width are greater than the sales plan amounts for that width, then the excess production is the total number of finished rolls minus the sales plan amount for that width. Naturally, you want as little excess production as possible.

If you set up the Calculations and Results section correctly, it should look like Figure 9-3.

		50" C	45" D	40" E	35" F	30" G	No. of Master Rolls Scheduled H	Scrap Wt per Master Roll I	Lbs Scrap per Layout J
13	**Calculations and Results Section:**								
15	Finished Roll Weight	500	450	400	350	300			
16	Finished Roll Widths	50"	45"	40"	35"	30"			
17	Line 41 (75" wide)	0	1	0	0	1	1	0	0
18	Possible Slitting Combinations	0	0	1	1	0	1	0	0
19		1	0	0	0	0	1	250	250
20		0	0	0	2	0	1	50	50
21		0	0	1	0	1	1	50	50
22		0	0	0	0	2	1	150	150
23	Total Master Rolls Line 41						6		
24	Line 42 (105" wide)	2	0	0	0	0	1	50	50
25	Possible Slitting Combinations	0	0	0	3	0	1	0	0
26		0	0	0	2	1	1	50	50
27		1	1	0	0	0	1	100	100
28		1	0	1	0	0	1	150	150
29		1	0	0	1	0	1	200	200
30		0	0	2	0	0	1	250	250
31		0	2	0	0	0	1	150	150
32		0	0	1	0	2	1	50	50
33		0	1	0	0	2	1	0	0
34		0	0	0	0	3	1	150	150
35	Total Master Rolls Line 42						11		
36								Total Scrap Weight lbs	
37	Finished Rolls by Width-Line 41	1	1	2	3	4			1650
38	Finished Rolls by Width-Line 42	5	4	4	6	8		RM Yield	
39	Total Finished Rolls by Width	6	5	6	9	12			89.7%
40	Sales Plan--Rolls by Width	1100	1230	3475	2800	5750			
42	Excess Production by Width	-1094	-1225	-3469	-2791	-5738			

FIGURE 9-3   Calculations and Results section with the formulas correctly entered

Cost Summary Section

The Cost Summary section contains calculations for the total pounds of raw material used by Lines 41 and 42, the machine hours needed on each line to complete the schedule, all of the manufacturing costs incurred by both lines, and a Totals column. See Figure 9-4. A description of each line item follows the figure.

	A	B	C	D	E
44					
45		**Cost Summary Section:**			
46			Line 41	Line 42	Totals
47		Total Lbs Raw Material Used			
48		Machine Hours Needed			
49		Raw Materials Cost			
50		Direct Labor Cost			
51		Mfg Overhead Cost			
52		Total Costs			

FIGURE 9-4   Cost Summary section

- Column headings—Enter the headings in cells C46 through E46 as shown.
- Total Lbs Raw Material Used—Cells C47 and D47 hold the total pounds of raw material used for Line 41. To calculate this value, multiply the total master rolls made on Line 41 (cell H23) by the master roll weight in pounds (cell C10). Similarly, calculate the total pounds of raw material used for Line 42 by multiplying the total master rolls made (cell H35) by the master roll weight in pounds (cell D10). Add the raw material weights for Lines 41 and 42 and place the resulting total in cell E47.
- Machine Hours Needed—Cells C48 and D48 hold the machine hours needed for Line 41. To calculate this value, divide the total pounds of raw material used for Line 41 (cell C47) by the production rate in pounds per hour (cell C9). Similarly, calculate the machine hours needed for Line 42 by dividing the total pounds of raw material used (cell D47) by the production rate in pounds per hour (cell D9).
- Raw Materials Cost—Cells C49 and D49 hold the cost of raw materials. To calculate this cost, multiply the total pounds of raw material used for Lines 41 and 42 (cells C47 and D47, respectively) by the raw material cost per pound in the Constants section. The cost per pound is

the same for both lines. Add the raw materials cost for both lines and place the resulting total in cell E49.

- Direct Labor Cost—Cells C50 and D50 hold the cost of direct labor. Calculate the cost for each line by multiplying the machine hours needed (cells C48 and D48) by the direct labor cost per machine hour in the Constants section (cells C7 and D7, respectively). Add the direct labor cost for both lines and place the resulting total in cell E50.
- Mfg Overhead Cost—Cells C51 and D51 hold the cost of manufacturing overhead. Calculate the cost for each line by multiplying the machine hours needed (cells C48 and D48) by the manufacturing overhead cost per machine hour in the Constants section (cells C8 and D8, respectively). Add the overhead cost for both lines and place the resulting total in cell E51.
- Total Costs—Cells C52 through E52 hold the total costs. The total cost for Line 41 in cell C52 is the sum of the raw materials cost, the direct labor cost, and the manufacturing overhead cost in cells C49 through C51. The total cost for Line 42 is determined in the same way as Line 41; enter this total in cell D52. Cell E52 holds the grand total of the manufacturing costs from Lines 41 and 42. This cell will be the optimization cell for your second Solver problem—it should be filled in the same color you used to fill the Total Scrap optimization cell. In the spreadsheet skeleton, cell E52 is filled in orange.

If you wrote the formulas correctly, the Cost Summary section values will look like those in Figure 9-5.

	A	B	C	D	E
44					
45		Cost Summary Section:			
46			Line 41	Line 42	Totals
47		Total Lbs Raw Material Used	4500	11550	16050
48		Machine Hours Needed	3	3.85	
49		Raw Materials Cost	$1,890	$4,851	$6,741
50		Direct Labor Cost	$111	$223	$334
51		Mfg Overhead Cost	$1,350	$3,176	$4,526
52		Total Costs	$3,351	$8,251	$11,602

FIGURE 9-5 Completed Cost Summary section

Your model is ready for scheduling production on Lines 41 and 42. Before applying Solver to determine the best slitting schedule, you should first try to schedule the production manually.

Scheduling Production Manually and Applying Solver to Minimize Total Scrap

Attempt to assign the slitting patterns manually in the changing cells; in other words, manually assign the number of master rolls for each slitting pattern for both lines. You have several good reasons for doing this. First, you can make sure your model is working correctly before you set up Solver to run. Second, assigning the schedule manually will demonstrate which constraints you must meet in solving the problem. For instance, you must meet or exceed the sales plan for the five different roll widths (cells C40 through G40), and the machine hours needed to complete the schedule cannot exceed the 2000 hours available for either Line 41 or Line 42.

Start by assigning values in the changing cells for the patterns that produce zero waste on both lines (cells H17 and H18 for Line 41 and cells H25 and H33 for Line 42). You will have to do a lot of juggling between the patterns to make a sufficient number of finished rolls and meet the sales plan. Eventually, you can produce a schedule that meets the sales plan, but note how much scrap weight (cell J36) and excess production in finished rolls (cells C42 through G42) you end up making. Figures 9-6 and 9-7 show a guess at a manual solution; it attempts to meet the sales plan for the larger-sized rolls (C38 and D38) as closely as possible, based on the assumption that the scrap and excess production weights will be minimized. Although the guess meets the constraints for the sales plan and machine hours, it is by no means an optimal solution—it produced more than 600,000 pounds of scrap and 200 excess 35" rolls (cells J36 and F42).

	A	B	C	D	E	F	G	H	I	J
12										
13		Calculations and Results Section:								
14										
15		Finished Roll Weight	500	450	400	350	300			
16		Finished Roll Widths	50"	45"	40"	35"	30"	No. of Master Rolls Scheduled	Scrap Wt per Master Roll	Lbs Scrap per Layout
17		Line 41 (75" wide)	0	1	0	0	1	0	0	0
18		Possible Slitting Combinations	0	0	1	1	0	300	0	0
19			1	0	0	0	0	0	250	0
20			0	0	0	2	0	0	50	0
21			0	0	1	0	1	0	50	0
22			0	0	0	0	2	1600	150	240000
23					Total Master Rolls Line 41			1900		
24		Line 42 (105" wide)	2	0	0	0	0	550	50	27500
25		Possible Slitting Combinations	0	0	0	3	0	900	0	0
26			0	0	0	2	1	0	50	0
27			1	1	0	0	0	0	100	0
28			1	0	1	0	0	0	150	0
29			1	0	0	1	0	0	200	0
30			0	0	2	0	0	1563	250	390750
31			0	2	0	0	0	0	150	0
32			0	0	1	0	2	50	50	2500
33			0	1	0	0	2	1230	0	0
34			0	0	0	0	3	0	150	0
35					Total Master Rolls Line 42			4293		
36									Total Scrap Weight lbs	660750
37		Finished Rolls by Width–Line 41	0	0	300	300	3200			
38		Finished Rolls by Width–Line 42	1100	1230	3176	2700	2560		RM Yield	88.86%
39		Total Finished Rolls by Width	1100	1230	3476	3000	5760			
40		Sales Plan--Rolls by Width	1100	1230	3475	2800	5750			
41										
42		Excess Production by Width	0	0	1	200	10			
43										

FIGURE 9-6 Manual solution, Calculations and Results section

	A	B	C	D	E
44					
45		Cost Summary Section:			
46			Line 41	Line 42	Totals
47		Total Lbs Raw Material Used	1425000	4507650	5932650
48		Machine Hours Needed	950	1502.55	
49		Raw Materials Cost	$598,500	$1,893,213	$2,491,713
50		Direct Labor Cost	$35,150	$87,148	$122,298
51		Mfg Overhead Cost	$427,500	$1,239,604	$1,667,104
52		Total Costs	$1,061,150	$3,219,965	$4,281,115

FIGURE 9-7 Manual solution, Cost Summary section

Name the solution worksheet **Premier Guess**, then right-click the worksheet name tab and click Move or Copy. The window shown in Figure 9-8 appears. Click the Create a copy check box to create a new worksheet named Premier Guess (2). Click the tab to see the new worksheet, then right-click the tab and rename the worksheet **Premier Min Scrap**. Use this worksheet to set up Solver to minimize total scrap.

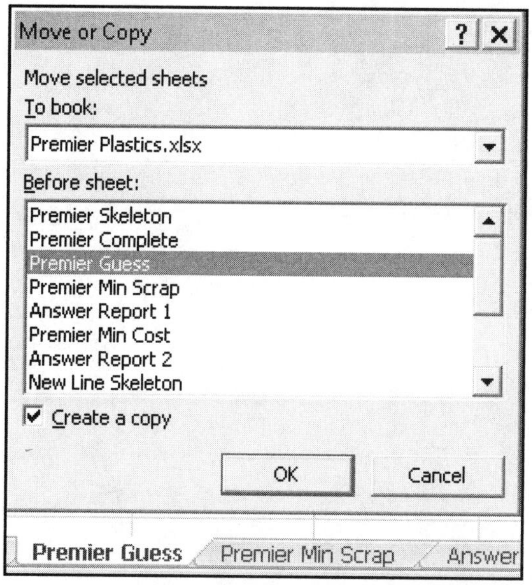

FIGURE 9-8 Copying the worksheet to a new tab

Setting up and Running Solver to Minimize Total Scrap

Before using the Solver Parameters window, you should jot down the parameters you must define and their cell addresses. Here is a suggested list:

- The cell you want to optimize (Total Scrap Weight lbs. in cell J36) and whether you want to minimize or maximize it
- The changing cells you want Solver to manipulate to obtain the optimal solution: No. of Master Rolls Scheduled, Line 41 and Line 42—cells H17 through H22 and H24 through H34
- The constraints you must define:
    - Each changing cell must be an integer that is greater than or equal to zero—you cannot make a fraction of a master roll.
    - The machine hours needed for each line from the Cost Summary section cannot be greater than the 2000 available machine hours listed in the Constants section.
    - The number of finished rolls of each of the five widths produced must be equal to or greater than the sales plan numbers.

As a reminder, Solver is in the Analysis group on the Data tab in Excel. Click Solver to open the Solver Parameters window, then enter the optimization cell in the Set Objective text box, click the Min button, and enter the changing cells. Remember that you must enter two ranges of cells: H17 through H22 for Line 41 and cells H24 through H34 for Line 42. Next, enter all of the constraints from the preceding list. You should be able to use the default Simplex LP method to solve the problem. Click the Options button to open the Options window (see Figure 9-9). Change the Integer Optimality to 1% from the default setting of 5%.

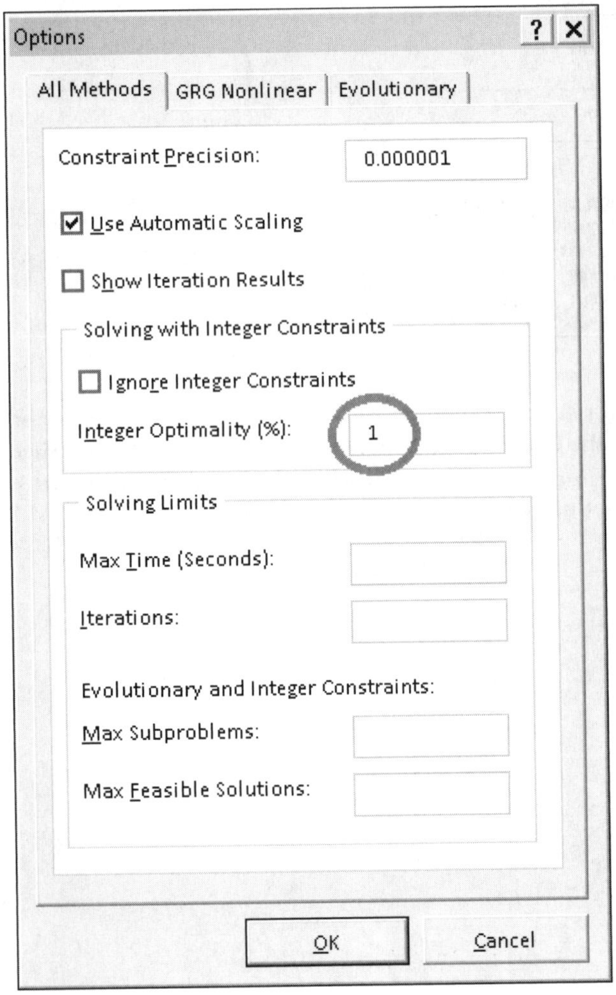

FIGURE 9-9 Integer Optimality set to 1%

Click OK to return to the Solver Parameters window, and then click Solve. Solver should find an optimal solution quickly. You want to keep the solution, but you also want an Answer Report, so click Answer under the Reports heading in the Solver Results window. See Figure 9-10.

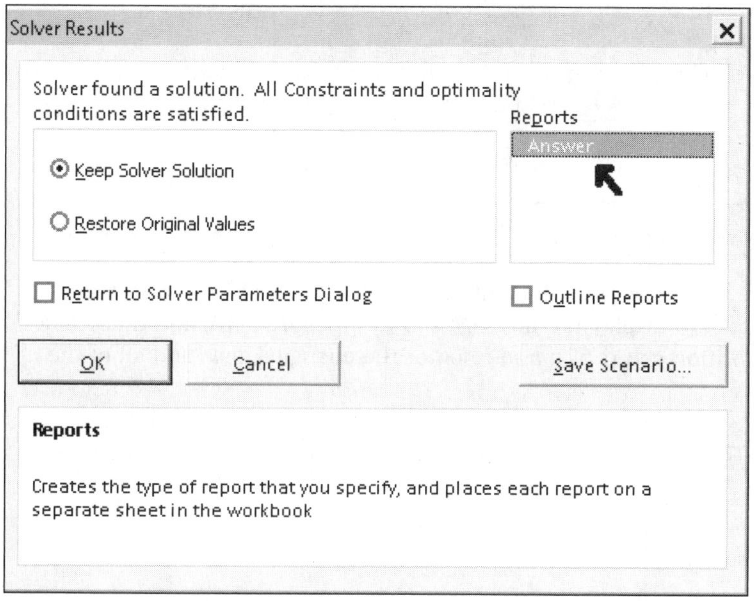

FIGURE 9-10   Creating an Answer Report

An Answer Report is useful because it reports both the original value in the optimization cell and the value that Solver achieved. Compare the two total scrap weights to see the scrap reduction achieved by using Solver.

Assignment 1B: Applying Solver to Minimize Total Manufacturing Cost

Compared with the manual guess, Solver probably reduced Premier's total scrap considerably while producing a feasible production schedule. However, a better solution is to minimize total manufacturing cost instead. The data in the Constants section shows that Line 42 yields twice the production rate in pounds per hour for less than twice the operating cost. If Solver is configured to minimize total manufacturing cost instead of total scrap, Premier might be able to use more of the available hours on Line 42 and save more money than with the minimized scrap solution.

Start by copying the Premier Min Scrap worksheet to a new worksheet, and then rename the new sheet **Premier Min Cost**. Open Solver and examine the Solver Parameters window carefully; the copied worksheet should contain the same Solver parameters as those in the original worksheet, but sometimes an error occurs and Solver tries to refer back to the cells in the original worksheet. If so, an error message appears, as shown in Figure 9-11.

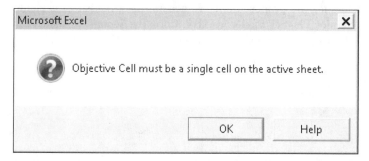

FIGURE 9-11   Solver erroneously attempts to link the cells back to the old worksheet

If you encounter this error, click OK. The Solver Parameters window will look like Figure 9-12.

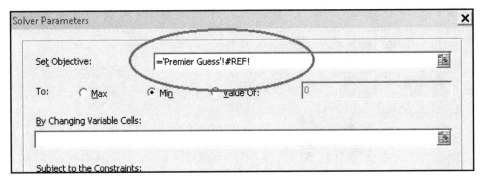

FIGURE 9-12   The Solver Parameters window with a reference error

Unfortunately, the only way to resolve this error is to clear all the values in the Solver Parameters window and enter them again. Click Reset All to display the message shown in Figure 9-13, and then click OK to clear the settings. Enter the new optimization cell (E52), and re-enter the changing cells and all of the constraints.

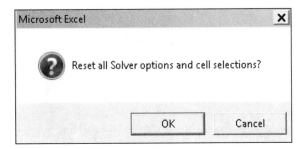

FIGURE 9-13   Resetting Solver

If the Solver Parameters window copied the previous parameters correctly, the only necessary modification is to change the optimization cell from total scrap (cell J36) to total manufacturing cost (cell E52). The other entries, including the changing cells, constraints, and solving method, should be correct. When you run Solver, it should find an optimal solution promptly. Create an Answer Report; Solver will name it Answer Report 2. If you set up Solver correctly to minimize total manufacturing cost, it should produce a schedule with a lower total cost than that in the Premier Min Scrap worksheet.

Assignment 1C: New Line Proposal

Fran is pleased with the scheduling model you built for Lines 41 and 42. She wants you to use a similar model to evaluate the possible savings from replacing the two older lines with a new line that can produce higher throughput and a master roll width of 120". You could modify the existing model for Lines 41 and 42 to create the model for the new line, but it is probably better to build a new decision model that creates an optimal schedule for the proposed line and performs a capital investment analysis on the potential savings.

Fortunately, a skeleton spreadsheet is available for the proposed line. To access the spreadsheet skeleton, locate Case 9 in the data files and then select **Premier Plastics Skeletons.xlsx**. Copy the second worksheet in the file, New Line Skeleton, for Assignment 1C. *It is strongly recommended that you use this skeleton to prepare your new model.*

As shown in Figure 9-14, the new worksheet is almost identical to the previous one you used, with two exceptions. The Constants section and Calculations and Results section are based on a 120" wide production line that has a much higher throughput capacity. The worksheet also includes a new Capital Investment section.

	A	B	C	D	E	F	G	H	I	J	K
1											
2		Premier Plastics--Packaging Film Production Scheduling--Proposed New Line									
3											
4		Constants Section									
5			New Line								
6		Raw Material Cost per pound	$0.42								
7		Direct Labor Cost per Mach	$65								
8		Mfg Overhead Cost per Mach	$1,000								
9		Production Rate Lbs/hr	5000								
10		Master Roll Wt in lbs	1200								
11		Available Machine Hours	2000								
12											
13		Calculations and Results Section									
14											
15		Finished Roll Weight (lbs.)	500	450	400	350	300				
16		Finished Roll Width	50"	45"	40"	35"	30"	Master Rolls Scheduled	Scrap Wt per Roll	Scrap per Layout	
17		New Line (120" wide)	0	0	3	0	0	0	0		
18		Possible Slitting Combinations	0	0	0	0	4	0	0		
19			0	0	0	3	0	0	150		
20			0	1	1	1	0	0	0		
21			2	0	0	0	0	0	200		
22			0	2	0	0	1	0	0		
23			1	0	1	0	1	0	0		
24			0	0	2	1	0	0	50		
25			1	0	0	2	0	0	0		
26			1	0	0	1	1	0	50		
27		Finished Rolls by Width--New Line							Total No. of Master Rolls		Total Scrap
28		Sales Plan--Rolls by Width	1100	1230	3475	2800	5750				
29		Excess Rolls Produced							RM Yield		
30											
31		Cost Summary Section:									
32			New Line								
33		Total Lbs Raw Material Used									
34		Machine Hrs needed									
35		Raw Materials Cost									
36		Direct Labor Cost									
37		Mfg Overhead Cost									
38		Total Costs									
39											
40											
41		Capital Investment Section--Proposed New Line									
42		Schedule of Cash Flows:									
43		Item	Timing	Amount							
44		Cost of New Line	End of Yr 0	($2,500,000)							
45		Total Mfg Cost, Old Lines	NA								
46		Total Mfg Cost, New Line	NA								
47		Annual Savings in Mfg Cost:	End of Yrs 1-6								
48		Residual Value (resell old lines)	End of Yr 2	$1,000,000							
49											
50		Cash Flow Stream	End of Yr	Amount							
51			0								
52			1								
53			2								
54			3								
55			4								
56			5								
57			6								
58			IRR								

FIGURE 9-14   Proposed New Line Skeleton worksheet

The worksheet for the proposed new line serves two purposes: It models the optimum production schedule for a 120" extrusion line, and more importantly, the Capital Investment section provides the analysis needed to help Premier make an informed decision about whether to invest in the new line.

If you successfully built the scheduling model for Lines 41 and 42, you should be able to complete the required formulas for the proposed new line. If you need help entering these formulas, refer to the model for Lines 41 and 42 or ask your instructor. Because the Capital Investment section and its associated formulas are new, a detailed explanation of the line items follows Figure 9-15.

	A	B	C	D
40				
41		**Capital Investment Section--Proposed New Line**		
42		**Schedule of Cash Flows:**		
43		Item	Timing	Amount
44		Cost of New Line	End of Yr 0	($2,500,000)
45		Total Mfg Cost, Old Lines	NA	
46		Total Mfg Cost, New Line	NA	
47		Annual Savings in Mfg Cost:	End of Yrs 1-6	
48		Residual Value (resell old lines)	End of Yr 2	$1,000,000
49				
50		Cash Flow Stream	End of Yr	Amount
51			0	
52			1	
53			2	
54			3	
55			4	
56			5	
57			6	
58			IRR	

FIGURE 9-15   Capital Investment section for proposed new line

If you used the skeleton spreadsheet to complete this part of the assignment, the Capital Investment section is labeled and formatted as shown in the figure. The Timing labels in column C provide a guide to entering the Cash Flow Stream values at the bottom of the section. Enter the following formulas and values:

- Cost of New Line—Cell D44 holds the amount of capital required to build the new line, which is estimated at $2.5 million. Note that the amount is a *negative* value. In Excel investment models, cash outflows are entered as negative values.
- Total Mfg Cost, Old Lines—Cell D45 holds the value of the minimized total manufacturing cost that Solver calculated in Assignment 1B. Enter this cost in the cell.
- Total Mfg Cost, New Line—Cell D46 holds the minimized total manufacturing cost that Solver will calculate for this worksheet. Enter "=C38" to transfer the total manufacturing cost for the new line from the Cost Summary section.
- Annual Savings in Mfg Cost—Cell D47 holds the annual savings in manufacturing costs. This value is the total manufacturing cost of the old lines (cell D45) minus the total manufacturing cost of the new line (cell D46).
- Residual Value (resell old lines)—Cell D48 holds the residual value of the old lines. Frequently, old manufacturing equipment that is still serviceable can be resold through equipment brokers. Premier's engineering staff has determined that the company can net $1 million from reselling the old lines, but the engineers estimate that it will take two years to find a buyer. In addition, Premier must keep the old lines in service until the new line is built and in production. Therefore, Premier will not recover the resell value until the end of Year 2.
- Cash Flow Stream—The values in cells D51 through D57 represent the cash outflows and inflows that will result if the new line is built. Enter the following values in these cells:
    - End of Yr 0—In cell D51, enter the cost of the new line from cell D44. Remember to make the value negative to represent a cash outflow.
    - End of Yr 1—In cell D52, enter the annual savings in manufacturing costs from cell D47. This value is a cash inflow, so it should be a positive number.
    - End of Yr 2—In cell D53, enter the sum of the annual savings from cell D47 and the residual value from selling the old lines from cell D48.
    - End of Yr 3 through 6—In cells D54 through D57, enter the annual savings in manufacturing costs from cell D47.
- IRR—Cell D58 holds the internal rate of return (IRR) for an investment with cash flows over a period of years. Premier Plastics requires an internal rate of return of 18% for major capital

projects. In other words, if the proposed new line does not return 18% or more over the six years of cash flows, Premier will not execute the project. To enter the IRR formula, click the Insert Function (f_x) button next to the formula bar below the Ribbon (see Figure 9-16). When the Insert Function window appears, enter "IRR" in the Search for a function text box. Click IRR when it appears in the Select a function text box, and then click OK. A parameters window appears, as shown in Figure 9-17. Enter "D51:D57" in the Values text box to enter the cash flow cells. Do not make an entry in the Guess text box. When you click OK, the IRR result appears in cell D58 as a percentage. The IRR result cell is filled in light blue.

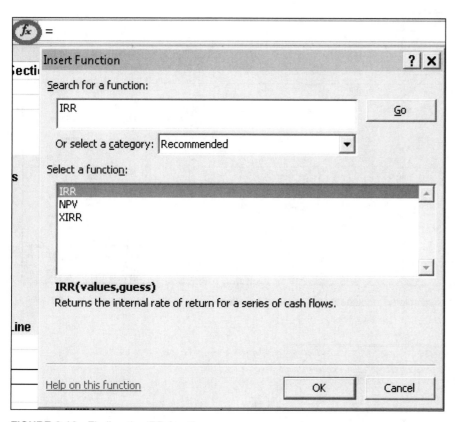

FIGURE 9-16   Finding the IRR function

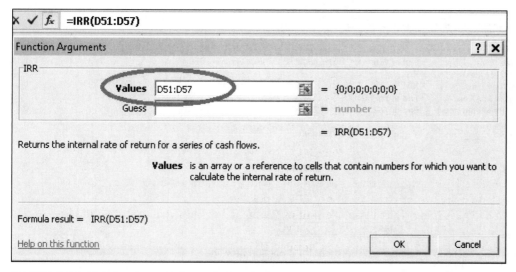

FIGURE 9-17   Entering the cash flow cells into the Values text box

If you set up the worksheet correctly for the proposed new line, the initial values should appear as shown in Figures 9-18 and 9-19. Note that Figure 9-19 does not include an entry in cell D45 for the total manufacturing cost of the old lines because doing so would reveal the solution from Assignment 1B. However, you should enter a value for this cell in the worksheet.

	A	B	C	D	E	F	G	H	I	J	K
1											
2		Premier Plastics--Packaging Film Production Scheduling--Proposed New Line--Formulas Completed									
3											
4		Constants Section:									
5			New Line								
6		Raw Material Cost per pound	$0.42								
7		Direct Labor Cost per Mach hour	$65								
8		Mfg Overhead Cost per Mach hour	$1,000								
9		Production Rate Lbs/hr	5000								
10		Master Roll Wt in lbs	1200								
11		Available Machine Hours	2000								
12											
13		Calculations and Results Section:									
14											
15		Finished Roll Weight (lbs.)	500	450	400	350	300				
16		Finished Roll Width	50"	45"	40"	35"	30"	No. of Master Rolls Scheduled	Scrap Wt per Roll	Lbs Scrap per Layout	
17		New Line (120" wide)	0	0	3	0	0	1	0	0	
18		Possible Slitting Combinations	0	0	0	0	4	1	0	0	
19			0	0	0	3	0	1	150	150	
20			0	1	1	1	0	1	0	0	
21			2	0	0	0	0	1	200	200	
22			0	2	0	0	1	1	0	0	
23			1	0	1	0	1	1	0	0	
24			0	0	2	1	0	1	50	50	
25			1	0	0	2	0	1	0	0	
26			1	0	0	1	1	1	50	50	
27		Finished Rolls by Width-New Line	5	3	7	8	7	Total No. of	10 Master Rolls	450	Total Scrap
28		Sales Plan--Rolls by Width	1100	1230	3475	2800	5750				
29		Excess Rolls Produced	-1095	-1227	-3468	-2792	-5743	RM Yield		96.25%	

FIGURE 9-18   Top two sections of completed worksheet for proposed new line

	Cost Summary Section:		
31	Cost Summary Section:		
32		New Line	
33	Total Lbs Raw Material Used	12000	
34	Machine Hrs needed	2.4	
35	Raw Materials Cost	$5,040	
36	Direct Labor Cost	$156	
37	Mfg Overhead Cost	$2,400	
38	Total Costs	$7,596	
39			
40			
41	Capital Investment Section--Proposed New Line		
42	Schedule of Cash Flows:		
43	Item	Timing	Amount
44	Cost of New Line	End of Yr 0	($2,500,000)
45	Total Mfg Cost, Old Lines	NA	
46	Total Mfg Cost, New Line	NA	$7,596
47	Annual Savings in Mfg Cost:	End of Yrs 1-6	($7,596)
48	Residual Value (resell old lines)	End of Yr 2	$1,000,000
49			
50	Cash Flow Stream	End of Yr	Amount
51		0	($2,500,000)
52		1	($7,596)
53		2	$992,404
54		3	($7,596)
55		4	($7,596)
56		5	($7,596)
57		6	($7,596)
58		IRR	-41.8%

FIGURE 9-19   Bottom two sections of completed worksheet for proposed new line

As you did with the original model, make a manual guess at a production schedule for the proposed new line. You should find it easier to create a good production schedule because six of the ten slitting patterns have zero scrap weight per master roll. Fran seems to have been correct in her opinion that a 120" line would better fit the finished roll widths. After you complete the manual schedule, name the worksheet **New Line Guess**. Next, make a copy of the worksheet and rename it **New Line—Min Cost**. Set up and run Solver to determine the optimal total manufacturing cost for the proposed new line. Create an Answer Report and compare your manual guess with the Solver solution. Finally, note the value for the internal rate of return. Did it meet or exceed Premier's requirement?

ASSIGNMENT 2: USING THE WORKBOOK FOR DECISION SUPPORT

You have built a series of worksheets, first to create an optimal production schedule for Lines 41 and 42, and then to create an optimal schedule for the proposed new line with an additional capital investment analysis. You will now complete the case by using your solutions and Answer Reports to gather the data needed for Premier to make decisions about production scheduling and capital investment. You will also document your recommendations in a memorandum.

Assignment 2A: Using Your Workbook to Gather Data

Print each worksheet and Answer Report. Print the worksheets in landscape orientation so that each fits on one page. Answer Reports are usually impossible to fit on one page, but your instructor may require you to print only the first page because it contains the original and optimized values for scrap pounds or total manufacturing cost. If your instructor requires that you print the entire report, resize the columns so the report fits in portrait orientation, and then print it horizontally.

Assignment 2B: Documenting Your Recommendations in a Memorandum

Use Microsoft Word to write a memo to the management team at Premier Plastics. Summarize the results of your analysis, include your recommendations regarding the optimal production schedule for the existing lines, and state whether you think the proposed new line meets Premier's capital investment requirements. Your memo should conform to the following requirements:

- Set up your memo as described in Tutorial E.
- In the first paragraph, briefly describe the situation and state the purpose of your analysis.
- Next, summarize the results of your analysis and state your recommendation for replacing the two older lines with the proposed 120" line. The recommendation should be based on the internal rate of return projected for the new line's investment.
- Support your recommendation with appropriate illustrations from your worksheets. To make the illustrations easier to read, consider using the Zoom function to make key sections of your worksheets appear larger. Then press the Print Screen key and paste the screenshots into the Microsoft Paint utility, where you can edit and crop. You can use Paint to create effective illustrations for your report. If you do not know how to use Paint, ask your instructor.

The capital investment analysis used in the worksheet for the proposed new line was simplified for this case. It addressed only cash flows and did not account for other financial considerations such as the following:

- The tax effect on reported income of the higher depreciation expense for a new line in its first six years of operation
- Tax credits offered to companies for improving manufacturing technology through capital investment
- Probabilities for success or failure of the project and expected values for cash flows to include resale value of the old equipment. The use of decision trees, probabilities, and expected values in capital investment decisions is referred to as Bayesian analysis.

Your instructor might provide guidance on how to address these issues in your report.

ASSIGNMENT 3: GIVING AN ORAL PRESENTATION

Your instructor may request that you summarize your analysis and recommendations in an oral presentation. If so, prepare a presentation for Fran Peters and other managers that lasts 10 minutes or less. When preparing your presentation, use Microsoft PowerPoint slides or handouts that you think are appropriate. Tutorial F explains how to prepare and give an effective oral presentation.

DELIVERABLES

Prepare the following deliverables for your instructor:

1. A printout of the memo
2. Printouts of all your worksheets and Answer Reports
3. Your Word document, Excel workbook, and PowerPoint presentation on electronic media or sent to your course site as directed by your instructor

Staple the printouts together with the memo on top. If you used more than one Excel workbook file for your case, write your instructor a note that describes the different files.

PART 4

DECISION SUPPORT CASE USING BASIC EXCEL FUNCTIONALITY

THE STATE PENSION FUNDING PROBLEM

Decision Support Using Excel

PREVIEW

Your state's pension fund has suffered investment losses in recent years. The governor has ordered a review of the situation to determine how the fund can maintain its assets in the next decade. In this case, you will use Microsoft Excel to help find a solution.

PREPARATION

- Review spreadsheet concepts discussed in class and in your textbook.
- Complete any exercises that your instructor assigns.
- Complete any parts of Tutorials C and D that your instructor assigns, or refer to them as necessary.
- Review the Microsoft Windows file-saving procedures in Tutorial C.
- Refer to Tutorials E and F as necessary.

BACKGROUND

Until recently, financial analysts thought your state's pension fund was very healthy. However, the fund has had more poor investment years than good ones since 2007, and the value of the fund's assets has fallen to $5 billion. The next decade will see increasing retirements, and state officials do not want the pension fund value to fall any lower. The governor has ordered a review of the fund's status and future prospects.

The pension fund gets money from three sources: (1) employee contributions, (2) state contributions, and (3) gains on its investments. The fund balance is decreased in three ways: (1) payments to pensioners, (2) payment of fund expenses, and (3) losses on investments. Each of these factors is discussed next.

- *Employee contributions*—All state employees are covered by the plan, and all must contribute part of their pay each year to the fund. Employees pay 3% of their gross pay over $6,000 into the fund. This 3% value is called the employee contribution factor. The $6,000 threshold is called a contribution shield. Both values are negotiated with the state employee labor union.
- *State contributions*—Your state has a simple formula: The state contributes 2.5 times what the employees contribute. The 2.5 coefficient is called the state factor. The value is negotiated with the state employee labor union.
- *Gains on investments*—Your state's pension fund manager invests in a mixture of stocks, U.S. government bonds, corporate bonds, and real estate. If the manager makes good decisions during a year, the value of the fund's assets will increase.
- *Payments to pensioners*—A pensioner's benefit is equal to 1.85% multiplied by the average of the employee's three highest yearly salaries, multiplied by the number of years the employee worked for the state. Then this benefit is increased by a cost-of-living adjustment, which currently is 3%. The 1.85% value is called the pension rate, and it is negotiated with the state employee union.

The number of pensioners is influenced by how many employees retire in a year and how many current pensioners die in a year. Baby boomers are reaching retirement age, and the state expects more retirements than usual for a few years. The state expects 7% of its pensioners to die each year.

- *Payment of fund expenses*—The state pays the fund manager to run the fund. In the coming year, the fee will be $5 million. The fee is expected to increase slightly each year.
- *Losses on investments*—If the fund's investment manager makes poor investment decisions in a year or investment markets have a poor year, the value of the fund's assets will decrease.

For the state, losing employees to retirement can be an opportunity to reduce costs. Some retiree positions can be eliminated altogether through automation or by reorganizing the work. For planning purposes, the state assumes that only 90% of retirees will be replaced. Thus, if 1,000 employees retire during a year, 900 new employees would be hired to replace them. The governor is determined to reduce the average employee salary by 1% each year. The goal is realistic, the governor thinks, because retirees are replaced by less expensive employees. Reducing total employee pay results in lower employee pension fund contributions, which in turn reduces the state pension fund contribution.

In the long run, the state fund manager says she expects an average annual 8% gain. Most investment managers say they use 8% for planning purposes, but average gains actually have been much less for most fund managers in recent years. With the high levels of uncertainty in today's investment markets, most financial managers no longer believe that 8% yearly gains are a reasonable expectation.

The governor has asked the state treasurer to assess the status of the state's pension fund. A ten-year assessment is required. The governor has a number of concerns.

1. She wants to know if the fund can maintain its asset value in the coming years.
2. Some state pension funds have been virtually bankrupted recently, and special supplemental state contributions have been needed. The governor wants to know if this scenario is likely for your state's fund.
3. The governor is determined to hold the line on employment costs, including payroll and pension costs. She wants to know how much money can be gained by challenging state employees to contribute more to the pension plan while allowing the state to contribute less. The employee would contribute more if the contribution shield was reduced from $6,000, exposing more of the salary to the 3% contribution factor. The employee also would contribute more if the contribution factor was raised above 3%. The state would contribute less if the cost-of-living adjustment was reduced from 3% or the state factor was reduced from 2.5.

The state treasurer has called you in to create a spreadsheet model of your state's pension fund to help answer the governor's questions.

ASSIGNMENT 1: CREATING A SPREADSHEET FOR DECISION SUPPORT

In this assignment, you will produce a spreadsheet that models the problem. Then, in Assignment 2, you will write a memorandum that explains your findings. In Assignment 3, you may be asked to prepare an oral presentation of your analysis.

A spreadsheet has been started, and is available for you to use. If you want to use the spreadsheet skeleton, locate Case 10 in your data files and then select **PensionFunding.xlsx**. Your worksheet should have the following sections:

- Constants
- Inputs
- Summary of Key Results
- Calculations

A discussion of each section follows.

Constants Section

Your spreadsheet should include the constants shown in Figure 10-1. An explanation of the line items follows the figure. Your model covers ten years, but values for only the first five are shown in Figure 10-1.

	A	B	C	D	E	F	G
1	**The State Pension Funding Problem**						
2							
3	**Constants**						
4	For all years:	Mortality rate:	7.00%	Rehire rate:	90%	Years on Job:	20
5		Pension rate:	1.85%	Pay reduction rate:	1%		
6	In each year:	**2012**	**2013**	**2014**	**2015**	**2016**	**2017**
7	Number of employees expected to retire in year	NA	2,000	2,500	3,000	3,000	3,000
8	Expected average of three highest salaries	NA	$50,000	$51,000	$52,000	$53,000	$52,000
9	Expected pension fund expenses	NA	$5,000,000	$5,050,000	$5,100,000	$5,150,000	$5,200,000

FIGURE 10-1   Constants section

- Mortality rate—Seven percent of retirees are expected to die each year.
- Rehire rate—Ninety percent of workers who retire in a year are expected to be replaced.
- Years on Job—On average, retirees work 20 years for the state.
- Pension rate—This percentage is multiplied by the average of the employee's three highest yearly salaries when computing the employee's pension payment.
- Pay reduction rate—Average employee pay will be 1% less each year.
- Number of employees expected to retire in year—The values shown are the number of retirees expected in each of the next ten years.
- Expected average of three highest salaries—The values shown are the expected average of retirees' three highest yearly salaries. This value is used when computing the employee's pension payment.
- Expected pension fund expenses—A total of $5 million will be paid to the pension fund manager in 2013. This amount is expected to increase slightly each year.

Inputs Section

Your spreadsheet should have the inputs shown in Figure 10-2. An explanation of the line items follows the figure.

	A	B	C	D	E	F
11	**Inputs**					
12	Expected earnings rate (.XX)		Employee			
13	Cost of living adjustment (.XX)		Contribution Shield		Employee	
14	Yearly State Supplemental Contribution		State Factor		Contribution Factor	

FIGURE 10-2   Inputs section

- Expected earnings rate—The fund manager uses 8%, but you want to be able to vary this rate as an input value.
- Cost of living adjustment—This factor is used in computing the pensioner payment. It is currently 3%, but you want to be able to vary it as an input value.
- Yearly State Supplemental Contribution—This value is an extra state contribution intended to increase the fund asset balance.
- Employee Contribution Shield—This factor is used in calculating the employee contribution. Currently the value is $6,000, but you want to be able to vary it as an input value.
- State Factor—This factor is used when computing the state's pension contribution. It is currently 2.5, but you want to be able to vary this factor as an input value.
- Employee Contribution Factor—This factor is used in calculating the employee contribution. It is currently 3%. You want to be able to vary this factor as an input value.

The cells should be formatted appropriately for numbers, currency, text, or percentages. Your instructor may require you to insert a comment and use conditional formatting in one or more cells. The existence of a comment is indicated by a diamond in the upper-right corner of a cell; the user places the mouse pointer over the diamond to see the comment. To enter a comment, right-click a cell and then choose Insert Comment from the menu. Conditional formatting is available from the Styles group on the Home tab; click the Conditional Formatting button and then click Highlight Cells Rules from the drop-down menu.

Summary of Key Results Section

Your worksheet should have the key results shown in Figure 10-3. The values are echoed from other parts of your spreadsheet. An explanation of the line items follows the figure. Your model covers ten years, but only the first five are shown in Figure 10-3 for Yearly surplus or deficit.

	A	B	C	D	E	F	G
16	**Summary of Key Results**	**2012**	**2013**	**2014**	**2015**	**2016**	**2017**
17	Yearly surplus or deficit	NA					
18	Accumulated surplus or deficit after tenth year	NA		NA	NA	NA	NA
19	Fund balance after ten years	NA		NA	NA	NA	NA

FIGURE 10-3 Summary of Key Results section

- Yearly surplus or deficit—This value is the increase or decrease in the pension fund balance during each year.
- Accumulated surplus or deficit after tenth year—This value is the change in fund balance over the ten-year period.
- Fund balance after ten years—This value is the pension fund balance after ten years.

Calculations Section

The first half of the Calculations section is shown in Figure 10-4. Values are calculated by formula, not hard-coded. Use cell addresses when referring to constants in formulas, unless otherwise directed. An explanation of the line items follows the figure. Your model covers ten years, but only the first five are shown in Figure 10-4.

	A	B	C	D	E	F	G
21	**Calculations**	**2012**	**2013**	**2014**	**2015**	**2016**	**2017**
22	Number of pensioners	24,000					
23	Number of new employees	NA					
24	Number of employees	42,000					
25	Average employee salary	$43,000					
26	Average payment to pensioner in year	NA					
27							
28	Contribution to pension fund:						
29	Employee Contribution	NA					
30	State Contribution	NA					
31	Supplemental State Contribution	NA					
32	Total Contribution	NA					

FIGURE 10-4 Calculations section

- Number of pensioners—This value is the number of pensioners receiving payments during the year. The value is a function of the prior year's value, the year's expected mortality rate, and the number of employees expected to retire in the year. To avoid data that includes a partial pensioner, use the INT() function or the ROUND() function with zero decimal places.
- Number of new employees—This value is a function of the number of employees expected to retire during the year and the rehire rate. To avoid data that includes a partial pensioner, use the INT() function or the ROUND() function with zero decimal places.
- Number of employees—This value is the number of state employees during the year. The value is a function of the number of employees last year, the number of retirees this year, and the number of new hires this year.
- Average employee salary—Average salary is expected to decline 1% per year. Thus, the 2013 average will be 1% less than the 2012 average, the 2014 average will be 1% less than the 2013 average, and so on.
- Average payment to pensioner in year—The base value is a function of the pension rate, the employees' expected average three highest yearly salaries, and the average retiree years on the job. After determining the base value, use it in the following formula:

$$\text{Base value} \times (1 + \text{cost-of-living adjustment})$$

The cost-of-living adjustment is from the Inputs section.

- Employee Contribution—This value is a function of the employee contribution factor from the Inputs section, the average employee salary reduced by the employee contribution shield from the Inputs section, and the number of employees during the year.
- State Contribution—This value is a function of the employee contribution from the Calculations section and the State Factor from the Inputs section.
- Supplemental State Contribution—This extra state contribution is a value from the Inputs section that should be echoed here.
- Total Contribution—This is the sum of the employee contribution, the regular state contribution, and the supplemental state contribution.

The second half of the Calculations section is shown in Figure 10-5. Here, total pension fund assets are computed for each year. Again, values are calculated by formula, not hard-coded. Use cell addresses when referring to constants in formulas, unless otherwise directed. An explanation of the line items follows the figure. Your model covers ten years, but only the first five are shown in Figure 10-5.

	A	B	C	D	E	F	G
34	Pension Fund Assets:						
35	Beginning Balance	NA					
36	Plus: Contribution to pension fund	NA					
37	Plus: Earnings on assets	NA					
38	Less: Payments to pensioners	NA					
39	Less: Pension fund expenses	NA					
40	Ending Balance	$5,000,000,000					
41							
42	Change in Pension Fund Balance:	2012	2013	2014	2015	2016	2017
43	In the year (surplus/deficit)	NA					
44	For ten years (surplus/deficit)	NA					

FIGURE 10-5 Calculations section, continued

- Beginning Balance—This value equals the pension fund asset balance at the end of the prior year.
- Plus: Contribution to pension fund—Total contributions to the pension fund is a calculated value that should be echoed here.
- Plus: Earnings on assets—This amount is a function of the expected earnings rate (from the Inputs section) and the beginning balance in the fund.
- Less: Payments to pensioners—This value is a function of the number of pensioners during the year and the average payment to pensioners during the year.
- Less: Pension fund expenses—This value is from the Constants section and can be echoed here.
- Ending Balance—This value equals the beginning balance plus the contributions and earnings, minus payments to pensioners and fund expenses.
- Change in Pension Fund Balance In the year (surplus/deficit)—This amount equals the ending fund balance for the current year minus the ending fund balance in the prior year. If the amount is positive, the fund had a surplus for the year. If the amount is negative, the fund had a deficit for the year.
- Change in Pension Fund Balance For ten years (surplus/deficit)—This amount equals the ending fund balance in the tenth year minus the fund balance at the beginning of the first year. In other words, the amount equals the sum of the yearly changes for the ten-year period. If the amount is positive, the fund had a surplus for the ten years. If the amount is negative, the fund had a deficit for the ten years.

ASSIGNMENT 2: USING THE SPREADSHEET FOR DECISION SUPPORT

You will now complete the case by (1) using the spreadsheet model to gather data needed to answer the governor's questions, (2) documenting your findings in a memo, and (3) giving an oral presentation if your instructor requires it.

Assignment 2A: Using the Spreadsheet to Gather Data

You have built the spreadsheet to develop "what-if" scenarios with the model's input values. Current input values are summarized in Figure 10-6.

Input	Current Value
Expected annual earnings rate	.08
Cost-of-living adjustment	.03
Yearly state supplemental contribution	$0
Employee contribution shield	$6,000
State factor	2.5
Employee contribution factor	.03

FIGURE 10-6   Current input values

The input values, except for the earnings rate, are the factors that the governor could try to change to improve the pension fund balance.

The governor wants you to investigate the following seven questions:

1. The 8% annual earnings rate is not expected, but if it is achieved, what will the fund balance be in ten years? What level of supplemental yearly contribution (if any) would be needed to maintain the fund balance at $5 billion? Manually enter values in the supplemental contribution cell until the fund balance is about $5 billion.

    You should rename a sheet as Question Answers. The sheet should include columns for the question number, fund balance, supplemental contribution, and comments. Enter the inputs and then manually record the results in the Question Answers sheet. The data is thus available in a convenient place when you write your memo later.

2. The governor thinks that a 6% average annual return is a reasonable expectation. What will the fund balance be in ten years with a 6% annual return? What level of supplemental yearly contribution (if any) would be needed to maintain the fund balance at $5 billion?

3. The governor is interested in knowing the worst-case scenarios. What would the fund balance be in ten years with a 0% annual return and no supplemental yearly contribution?

4. Continuing the interest in worst-case scenarios, what annual rate of return would be needed to get a zero fund balance after ten years (with no supplemental yearly contribution)? To see this scenario, manually enter values in the annual return input cell until the fund balance is about $0. Negative rates are possible. Enter the rate as a comment in the Question Answers sheet.

5. What would the governor have to negotiate with the employee union and the legislature to change the input factors? The governor thinks that a realistic outcome would be the set of factors shown in Figure 10-7.

Input	Negotiated Value
Cost-of-living adjustment	.01
Yearly state supplemental contribution	$0
Employee contribution shield	$3,000
State factor	2.0
Employee contribution factor	.0325

FIGURE 10-7   Realistic negotiated factors

Employees would contribute more because the contribution shield would be halved and the employee contribution rate would increase from .03. The state would contribute less because the state factor would decrease from 2.5 and the cost-of-living adjustment would decrease from .030.

Using the realistic negotiated factors, what will the fund balance be in ten years with a 6% annual return? What level of supplemental yearly contribution (if any) would be needed to maintain the fund balance at $5 billion?

6. Using the realistic negotiated factors, what will the fund balance be in ten years with a pessimistic 4% annual return? What level of supplemental yearly contribution (if any) would be needed to maintain the fund balance at $5 billion?

7. The governor envisions "hard-nosed" negotiations with the employee union and the legislature to change the input factors. These factors are shown in Figure 10-8.

Input	Negotiated Value
Cost-of-living adjustment	.0
Yearly state supplemental contribution	$0
Employee contribution shield	$0
State factor	2.0
Employee contribution factor	.0350

FIGURE 10-8 "Hard-nosed" negotiated factors

The governor thinks that the hard-nosed factors would be the best-case scenario for the state. Using these factors and different rates of return, what supplemental contribution would be needed to maintain a $5 billion fund? Rename a sheet as Hard Nosed. Use expected annual earnings rates from 6% to 8% with quarter-percent increments. Summarize the earnings rates and required state supplemental contributions in two columns. Portray the results in a chart, as shown in Figure 10-9 (data are illustrative only).

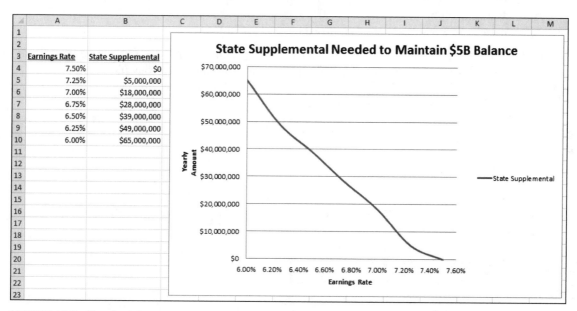

FIGURE 10-9 Required "hard-nosed" supplemental contributions

To create an Excel chart, first highlight the data set, including the labels; next, click the Insert tab and then choose the type of chart you want from the Charts group. (Figure 10-9 shows a scatter chart with smooth lines.) In the Labels group of the Layout tab, use the Chart Title and Axis Titles buttons to create titles and labels. To adjust the range of values for an axis, double-click the axis to open the Format Axis window, and then enter values for the Minimum and Maximum Axis Options.

When you finish gathering data for the seven questions, print the model's worksheets with any set of inputs. Then save the spreadsheet one last time.

Assignment 2B: Documenting Your Findings and Recommendation in a Memorandum

Now you will document your findings in a memo that answers the seven questions from the preceding section. Use the following tips to help prepare your memo in Word:

- Your memo should have proper headings such as Date, To, From, and Subject. You can address the memo to the governor and set it up as discussed in Tutorial E.
- Briefly outline the situation. However, you need not provide much background—you can assume that readers are familiar with the situation.
- List and then answer the seven questions in the body of the memo.
- What general conclusion can you draw about the status of the pension plan, now that your analysis is completed? What are the chances of keeping the $5 billion balance in the next ten years without significant supplemental contributions? What are the chances that the fund will go broke in the next ten years? State your conclusions, referring to the data as support.
- Include tables and/or charts to support your claims, as your instructor specifies. Tutorial E explains the procedures for creating a table in Word.

ASSIGNMENT 3: GIVING AN ORAL PRESENTATION

Your instructor may request that you summarize your analysis and results to the governor's staff in an oral presentation. Prepare to talk to the group for 10 minutes or less. Use visual aids or handouts that you think are appropriate. Tutorial F explains how to prepare and give an oral presentation.

DELIVERABLES

Assemble the following deliverables for your instructor:

1. Printout of your memo
2. Spreadsheet printouts
3. Flash drive or CD that includes your Word file and Excel file

Staple the printouts together with the memo on top. If you have more than one .xlsx file on your flash drive or CD, write your instructor a note that identifies your spreadsheet model's .xlsx file.

PART 5

INTEGRATION CASES USING ACCESS AND EXCEL

CASE **11**

THE COMMON STOCK ANALYSIS

Decision Support with Access and Excel

PREVIEW

Technical analysis and fundamental analysis are the two main approaches to analyzing the value of a common stock. Technical analysis uses the stock's historical price movement to predict future prices. Fundamental analysis uses data in a company's financial statements to assess the stock's value. The two methods can be used together. In this case, you will use Microsoft Access and Excel to analyze seven common stocks and determine if any are worth buying.

PREPARATION

- Review database and spreadsheet concepts discussed in class and in your textbook.
- Complete any exercises that your instructor assigns.
- Complete any parts of Tutorials B, C, and D that your instructor assigns, or refer to them as necessary.
- Review file-saving procedures for Microsoft Windows programs, as discussed in Tutorial C.
- Refer to Tutorials E or F as necessary.

BACKGROUND

You are thinking about buying shares of a company's common stock in the stock market. The stock is currently priced at $30. Is the stock price likely to go up or down in the future?

Two kinds of analysis can help you decide if a stock price is a good value: fundamental analysis and technical analysis.

In technical analysis, you study the pattern of a company's stock prices. You can perform technical analysis in many ways; some methods are basic and some are very complex.

In fundamental analysis, you use data in a company's balance sheet and income statement to assess the company's common stock value, which may differ from the current market price. As with technical analysis, you have many options for performing fundamental analysis; these options range from basic to very complex.

Technical analysts assume that all you need to know about a company is reflected in the stock's price, but fundamental analysts believe that better information comes from assessing the company's financial data. The two methods are not mutually exclusive, however. An analyst might gain more confidence in a stock if the same "buy" or "sell" recommendation results from each analysis.

A wealthy friend has asked you to analyze seven common stocks with code names A to G. Your friend does not own any shares of these companies, but will buy shares that you recommend. Data for the seven companies is in a file called **LikelyStocks.accdb**, which is available to you.

You decide to use the basic technical and fundamental analysis techniques discussed in the next sections.

Basic Fundamental Analysis

A company's income statement contains its revenues (sales), expenses, and net income. Net income is the difference between revenues and expenses. Typically, an income statement contains data from two or more years. A fundamental ratio is return on sales (ROS), which is the ratio of net income to revenue. The higher the ROS value is, the better. Usually, the current year's ratio is compared to the previous year's ratio. An improving ratio suggests improvement in a company's operations.

A company's balance sheet shows the productive assets owned by the company, its liabilities, and the stockholder's equity, which equals assets minus liabilities. A fundamental ratio is return on assets (ROA), which is the ratio of net income to total assets. The higher the ROA value is, the better. Usually, the current year's ratio is compared to the previous year's ratio. An improving ratio suggests improvement in a company's financial standing.

The LikelyStocks.accdb file contains tables that summarize financial data for the seven companies. Figure 11-1 shows the data for the current year; values are in billions of dollars.

ThisYearFinancial			
Company	Revenue	Net Income	Assets
A	$21.00	$0.25	$39.30
B	$134.20	-$3.60	$2,260.00
C	$199.30	$19.00	$184.80
D	$35.10	$11.80	$72.90
E	$150.20	$11.60	$751.20
F	$383.20	$30.50	$302.50
G	$31.50	$3.00	$40.40

FIGURE 11-1   Financial data for current year

The file contains a companion table that shows prior-year financial data of the same sort for the seven companies.

For each of the seven companies, you will compute the ROA and ROS ratios for the current year and prior year. You will consider buying stock in companies that have improving ROA and ROS ratios.

Basic Technical Analysis

Technical analysis uses a series of common stock prices. Typically, the data is charted for ease of analysis.

The LikelyStocks.accdb file contains tables that summarize stock prices for the seven companies. Figure 11-2 shows the data for Company A.

Company A	
Quote#	Close
1	$10.26
2	$10.15
3	$10.17
4	$9.69
5	$9.10

FIGURE 11-2   Company A stock prices

The file contains corresponding tables for companies B through G. Each table has approximately 500 quotes. A price at the end of a trading day on the stock exchange is called the "close." Table A has more than 500 closing prices. Each price is numbered in the table: Quote # 1 is the oldest listed quote from recent trading, Quote # 2 is the second oldest quote, and so on. The last record in the table is the most recent closing market quote for the stock.

For most companies, day-to-day stock prices can fluctuate greatly. This variation is sometimes referred to as "noise" in the data. The variation can make it difficult to see the underlying "real" trend in the stock's price movement. A hypothetical example is shown in Figure 11-3, which illustrates a company's closing stock price for 20 trading days on the exchange. Day 1 is the oldest price, and day 20 shows the current price.

	A	B
1	Day	Quote
2	1	$43.58
3	2	$43.37
4	3	$47.18
5	4	$45.82
6	5	$45.68
7	6	$47.04
8	7	$49.70
9	8	$51.17
10	9	$50.87
11	10	$50.65
12	11	$55.08
13	12	$55.08
14	13	$55.95
15	14	$58.72
16	15	$59.02
17	16	$61.40
18	17	$59.87
19	18	$59.16
20	19	$64.07
21	20	$65.67

FIGURE 11-3   Twenty days of stock prices for a company

The variation in the data series is shown in the chart in Figure 11-4.

FIGURE 11-4   Chart of stock prices

One way to dampen the variation and see the basic trend more clearly is to compute a "moving average" of the data, and then analyze the trend in the average. A simple moving average (MA) sums all the closing prices over the time period and divides the result by the number of prices. For example, in a five-day moving average, the five most recent closing prices are summed and then divided by five.

In effect, the raw data series is converted to a series of moving averages. Figure 11-5 continues the example by showing 10- and five-day moving averages.

D11			f_x	=SUM(B2:B11)/10	
	A	B	C	D	E
1	Day	Quote			
2	1	$43.58			
3	2	$43.37			
4	3	$47.18			
5	4	$45.82			
6	5	$45.68			
7	6	$47.04			
8	7	$49.70			
9	8	$51.17			
10	9	$50.87	Day	10 Day MA	5 Day MA
11	10	$50.65	1	$47.51	$49.89
12	11	$55.08	2	$48.66	$51.50
13	12	$55.08	3	$49.83	$52.57
14	13	$55.95	4	$50.70	$53.53
15	14	$58.72	5	$51.99	$55.10
16	15	$59.02	6	$53.33	$56.77
17	16	$61.40	7	$54.76	$58.03
18	17	$59.87	8	$55.78	$58.99
19	18	$59.16	9	$56.58	$59.63
20	19	$64.07	10	$57.90	$60.70
21	20	$65.67	11	$59.40	$62.03

FIGURE 11-5   Ten- and five-day moving averages

The figure shows the formula in cell D11, which computes the 10-day moving average for days 1 to 10 in the series. Notice that the formula adds the quotes in cells B2..B11 and then divides by 10. Can you imagine what the formula would be for the second day in the 10-day moving average series? The formula is shown in Figure 11-6.

D12			f_x	=SUM(B3:B12)/10	
	A	B	C	D	E
1	Day	Quote			
2	1	$43.58			
3	2	$43.37			
4	3	$47.18			
5	4	$45.82			
6	5	$45.68			
7	6	$47.04			
8	7	$49.70			
9	8	$51.17			
10	9	$50.87	Day	10 Day MA	5 Day MA
11	10	$50.65	1	$47.51	$49.89
12	11	$55.08	2	$48.66	$51.50

FIGURE 11-6   Formula for second 10-day moving average

As you can see, the first day's quote drops out of the sum and the 11th day's quote moves in. You can copy the formula from cell D11 to D12 to get this calculation.

The formula for the first five-day moving average is shown in Figure 11-7.

	E11	▼		f_x	=SUM(B7:B11)/5
	A	B	C	D	E
1	**Day**	**Quote**			
2	1	$43.58			
3	2	$43.37			
4	3	$47.18			
5	4	$45.82			
6	5	$45.68			
7	6	$47.04			
8	7	$49.70			
9	8	$51.17			
10	9	$50.87	**Day**	**10 Day MA**	**5 Day MA**
11	10	$50.65	1	$47.51	$49.89
12	11	$55.08	2	$48.66	$51.50

FIGURE 11-7   Formula for first five-day moving average

Here, five quotes are summed, and the result is divided by five. Note that this moving average is shown as the value that corresponds to the first day in the 10-day moving average.

The 10-day moving average is charted in Figure 11-8.

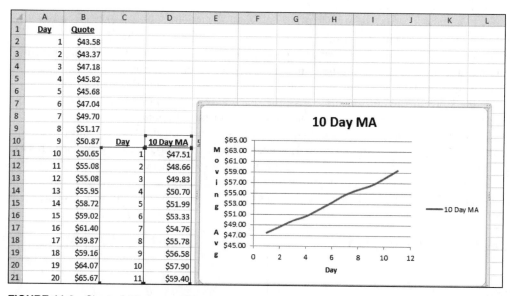

FIGURE 11-8   Chart of 10-day moving average

Clearly, this is a smoother trend line than the raw-data trend line shown in Figure 11-4. Technical analysts think that the moving average trend line is a truer representation of the basic trend than the raw-data trend line.

Figure 11-9 compares the 10-day and five-day trend lines.

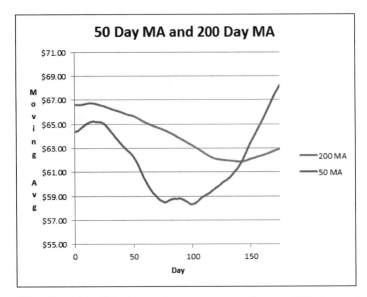

Day	Quote		Day	10 Day MA	5 Day MA
1	$43.58				
2	$43.37				
3	$47.18				
4	$45.82				
5	$45.68				
6	$47.04				
7	$49.70				
8	$51.17				
9	$50.87		1	$47.51	$49.89
10	$50.65		2	$48.66	$51.50
11	$55.08		3	$49.83	$52.57
12	$55.08		4	$50.70	$53.53
13	$55.95		5	$51.99	$55.10
14	$58.72		6	$53.33	$56.77
15	$59.02		7	$54.76	$58.03
16	$61.40		8	$55.78	$58.99
17	$59.87		9	$56.58	$59.63
18	$59.16		10	$57.90	$60.70
19	$64.07		11	$59.40	$62.03
20	$65.67				

FIGURE 11-9   Chart of 10-day and five-day moving averages

Why is the five-day line above the 10-day line? The overall trend is up, which means that the average of the five most recent values generally will exceed the average of the ten most recent values.

Technical analysts sometimes compare two moving averages—for example, 50-day and 200-day moving averages. An example will illustrate why. Assume that an analyst has a 200-day moving average for a stock. If the analyst can provide evidence that the stock's price will climb and keep climbing, the analyst will recommend purchasing the stock at the current price. Assume that today's stock price is greater than the current value of the 200-day moving average. Would that one data point be sufficient evidence that a sustained upward price trend is at hand? Probably not—today's rise might be mostly "noise," and the price might fall below the 200-day moving average tomorrow. However, what if the 50-day moving average crossed over the 200-day moving average? That crossover could be good evidence of a sustained upward shift in the trend, and a signal that the stock's shares should be purchased now.

Figure 11-10 is a chart of a 50-day moving average that has crossed over a 200-day moving average.

FIGURE 11-10   Fifty-day moving average crossing over 200-day moving average

In the chart, data is illustrative only; approximately 175 data points are shown. You can see that the 50-day average crossed over the 200-day moving average about 25 days ago. (Stock exchanges are not open every day of the year.) Analysts think that a 200-day moving average is a good measure of a trading year's trend, and that a 50-day moving average is a good measure of a quarterly trend. So, if quarterly averages are now above yearly averages, a major trend shift may be at hand, and it may be time to buy.

By the same logic, if quarterly averages fall below yearly averages, a major trend shift may be indicating that it is time to sell. Figure 11-11 is a chart of a 50-day moving average that has crossed under a 200-day moving average.

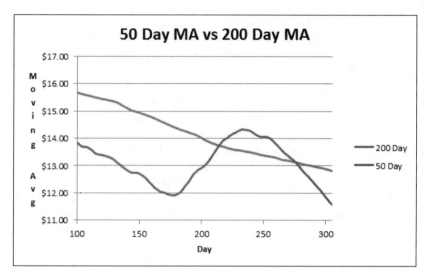

FIGURE 11-11   Fifty-day moving average crossing under 200-day moving average

In the chart, data is illustrative only. You can see that the 50-day average crossed under the 200-day moving average about 25 days ago, which suggests that the price will continue to fall and that the shares should be sold.

Your friend has asked you to analyze seven stocks. You will recommend buying stock in companies (1) whose 50-day moving averages have crossed over the 200-day moving averages *and* (2) that have improving ROA and ROS ratios.

A database file is available for you to use. To get the file, locate Case 11 in your data files and then select **LikelyStocks.accdb.**

ASSIGNMENT 1: MAKING QUERIES IN ACCESS

In this assignment, you will design and run two queries in Access.

ThisYearRatios Query

Create a query that computes and shows the return on sales and return on assets ratios for the seven companies based on this year's financial data. Your output should look like that in Figure 11-12; only the first two of the seven records are shown. Name the query ThisYearRatios.

ThisYearRatios		
Company ▾	ThisYearROS ▾	ThisYearROA ▾
A	0.012	0.006
B	-0.027	-0.002

FIGURE 11-12   ThisYearRatios query output

Note that ratios are formatted for three decimal places. To set this format, highlight the output column and select Properties from the menu. Select Standard number Format, three Decimals.

LastYearRatios Query

Create a query that computes and shows the return on sales and return on assets ratios for the seven companies based on last year's financial data. The output should have the same format shown in Figure 11-12. Name the query LastYearRatios.

When you finish the queries, save and close the **LikelyStocks.accdb** file.

ASSIGNMENT 2: USING EXCEL FOR DECISION SUPPORT

In this assignment, you will import the two Access queries into Excel and import quote data for the seven companies into Excel. You will create 50- and 200-day moving average charts for each company. You will then develop information needed to decide which stocks should be purchased.

Importing Quote Data Tables

Open a new file in Excel and save it as **RatioAnalysis.xlsx**. Rename Sheet1 as Company A. From the LikelyStocks database file, import the Company A table records into the Company A worksheet. Click the Data tab, select Get External Data, and then select From Access. Specify the Access filename, the table's name, and where to place the data in Excel (cell A1 is recommended).

The data will be imported into Excel as an Excel data table. In the Tools group, select Convert to Range. The data table's fill colors will remain. To change to the standard white fill, select an empty cell, click the Format Painter button in the clipboard group, and then highlight the data range. Finally, highlight the Close data and change it to Currency format.

By formula, compute the 50- and 200-day moving averages. Create an XY Scatter chart that shows the two moving averages. To make the chart, first highlight the data, including the labels. Next, select the Insert tab and choose the type of chart you want from the Charts group. Also in the Charts group, select Other Charts—All Chart Types, and then select XY (Scatter)—Scatter with Smooth Lines. In the Labels group of the Layout tab, use the Chart Title and Axis Titles buttons to create titles and labels. To adjust the range of axis values, double-click an axis to open the Format Axis window, and then enter values for the Minimum and Maximum Axis Options.

Generate a new worksheet, name it Company B, and then repeat the same process for Company B that you used to create Company A's chart. Repeat this process for all companies through G. When you finish, you will have seven worksheets with moving average charts.

Importing Queries

Generate a new worksheet and name it Fundamentals. Then import the two queries into the worksheet, side by side. Again, you will need to change data tables to data ranges, and you might need to change numerical formats to three decimal places. Delete the redundant Company column. When you finish, the upper part of the worksheet should look like Figure 11-13; only data for companies A and B are shown.

	A	B	C	D	E	F
1	Company	LastYearROS	LastYearROA		ThisYearROS	ThisYearROA
2	A	-0.063	-0.030		0.012	0.006
3	B	-0.015	-0.001		-0.027	-0.002

FIGURE 11-13   Imported query output

Using Ratio Data and Charts to Gather Information for Decision Making

You now want to use the ratio data and charts to gather information needed to decide on the seven stocks.

For each company, indicate improvement in the ROS ratios. Use an IF statement to compute the improvement. This part of the worksheet could be organized as shown in Figure 11-14; only Company A is shown.

	A	B	C
10	Return on Sales Analysis		
11			This Yr > Last Yr?
12		A	YES

FIGURE 11-14   Return on sales ratio improvement

You should create a similar section to summarize the ROA ratios.

Inspect each moving average chart. Note whether the 50-day MA has crossed over the 200-day MA and remains above it. You will follow this rule: If the 50-day MA has crossed over the 200-day MA after the 100-day mark, and remains above the 200-day MA, you will consider this stock for purchase. In a section of the Fundamentals worksheet, summarize your inspection as shown in Figure 11-15. Only Company A results are shown.

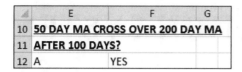

	E	F	G
10	**50 DAY MA CROSS OVER 200 DAY MA**		
11	**AFTER 100 DAYS?**		
12	A	YES	

FIGURE 11-15 Crossover inspection results

Note that you merely enter "YES" or "NO"; no formula is needed.

You should also note the shape of the curves when the 50-day MA is above the 200-day MA. Is the 50-day MA line trending down toward the 200-day MA line, indicating that it may cross back over in the near future? Make a side note for later reference for each of these possible reversals.

Devote a section of your spreadsheet to applying your investment decision rule. The logic for a company is: If the ROS ratio and ROA ratio have both improved, and if the 50-day MA has crossed over the 200-day MA, then you recommend investing in the stock. Otherwise, you do not recommend investing in the stock. Use an IF statement to indicate a "YES" or "NO" decision for each stock. Also, for later discussion with your friend, indicate whether a reversal is possible—perhaps your friend would not want to invest in such a stock. Your summary section should look like Figure 11-16; only the results for Company A are shown.

	E	F	G	H	
20	**Invest in this stock?**				
21		**Logic**		**Comment**	
22	A	YES		Reversal possible	

FIGURE 11-16 Summary of results

ASSIGNMENT 3: DOCUMENTING FINDINGS IN A MEMORANDUM

In this assignment, you write a memorandum in Microsoft Word that documents your findings and recommendations. You should describe the investment situation briefly. Your memo should then state which stocks you would and would not recommend.

In your memo, observe the following requirements:

- Your memo should have proper headings such as Date, To, From, and Subject. You can address the memo to your friend. Set up the memo as discussed in Tutorial E.
- Briefly outline the situation. However, you need not provide much background—you can assume that your friend is familiar with the situation.
- State your findings and recommendations in the body of the memo.
- Support your work by showing recommendations in a table. Tutorial E describes how to create a table in Word.

Your instructor might also request that you summarize your analysis and results to your friend in an oral presentation. If so, prepare to talk to your friend for 10 minutes or less. Use visual aids or handouts that you think are appropriate. Tutorial F explains how to prepare and give an oral presentation.

DELIVERABLES

Assemble the following deliverables for your instructor:

1. Printout of your memo
2. Spreadsheet and database printouts, as required by your instructor
3. Electronic media such as flash drive or CD, which should include your Word file, Access file, and Excel file

Staple the printouts together with the memo on top. If you have more than one .xlsx file or .accdb file on your flash drive or CD, write your instructor a note that identifies the files in this assignment.

THE TEACHER EVALUATION SYSTEM

Decision Support with Access and Excel

PREVIEW

Your school district wants to evaluate teachers based on their students' scores on standardized tests. In this case, you will use Microsoft Access and Microsoft Excel to create the evaluation system.

PREPARATION

- Review database and spreadsheet concepts discussed in class and in your textbook.
- Complete any exercises that your instructor assigns.
- Complete any parts of Tutorials B, C, and D that your instructor assigns, or refer to them as necessary.
- Review file-saving procedures for Windows programs, as discussed in Tutorial C.
- Refer to Tutorials E and F as necessary.

BACKGROUND

Your state's education officials complain that virtually all the teachers in your school district are rated as good or excellent by their school principals. There are surely bad teachers, state officials say, but they are not identified and the problem is not addressed.

Therefore, the public school district has been instructed by state education officials to devise an accountability system to identify good teachers and bad teachers. Good teachers would be paid more. Bad teachers would be retrained so that they improve, or they would be fired. Fired teachers then would be replaced by good teachers. In this way, officials believe that teaching would improve and students would learn more.

In the new system, subjective evaluations by school principals will not be the only measure of teacher success. In addition to principal evaluations, an objective method is needed to measure teacher performance. Your state's standardized tests are considered a reliable measure of student learning, which means that the test scores could be used to help measure teacher performance objectively. According to some state officials, teachers whose students do well on the tests are doing a good job; poor test scores must be a sign of bad teaching. The concept of matching teacher performance with student performance is highly controversial.

In your state, public school begins with kindergarten. Starting in first grade, elementary students take a standardized test at the end of each school year. The test has a reading section and a math section; the summary score ranges from 0 to 1000. These standardized tests are given for grades 1 through 8; the test becomes more difficult and comprehensive each school year.

Your district superintendent thinks that parental involvement is a key factor in a student's motivation and learning. The superintendent wants parents to supervise their children's homework, meet regularly with teachers, and be involved with the school. The superintendent thinks that children of "involved" parents usually do better in school. In your district, K–8 teachers are required to note the extent of parental involvement, and to report this information as a rating to the principal and the superintendent. Teachers are required to document these ratings by recording missed meetings, missed homework assignments, and so on. District officials think the ratings are reliable because they are documented and are rarely debated by parents.

You have been asked to develop an accountability system that incorporates standardized test scores and parental involvement ratings. Your model will be a prototype. The model will use test score data for all second-grade students in the district. To develop a sample group for comparison, you have identified three second-grade classes in a district school. You have two years of data for students in these classes.

An Access database called **TeacherEval.accdb** contains your data. The database is available for you to use. To get the file, go to your data files, select Case 12, and then select **TeacherEval.accdb**.

The tables in the file are discussed next. Figure 12-1 shows the first few records of the Year1Students table.

Year1Students	
StudentNum ▾	LackOfInvolvement? ▾
1	YES
2	YES
3	NO

FIGURE 12-1   Year1Students table records

The table shows each student's ID number and indicates whether parents are involved in the student's education. Each second-grader in the district is assigned a unique student number. A lack of parental involvement is indicated by the text value "YES" in the LackOfInvolvement? field. If the parents *are* involved, the entry is "NO."

The district had 548 second-grade students in Year 1 of your model. A companion table called Year2Students holds data for second-grade students in Year 2 of your model. The two tables contain two different sets of students. By sheer coincidence, Years 1 and 2 had the same number of second-grade students, 548.

Your model will also include two years of student records for three second-grade teachers in a particular district school. The teachers are named Smith, Jones, and Casey. Figure 12-2 shows the first few records of the Year1ClassAssignments table.

Year1ClassAssignments		
StudentNum ▾	ClassNum ▾	Teacher ▾
1	1	Smith
2	1	Smith
5	3	Casey
10	1	Smith

FIGURE 12-2   Year1ClassAssignments table records

The three teachers each had 23 students in their second-grade class each year. Each teacher's class has a unique number. Here, teacher Smith's Year 1 class was number 1. The table shows how individual students were assigned to each class. Each student is given a unique number, which is the table's key field. Figure 12-2 shows that teacher Smith had students 1, 2, and 10 among the 23 assigned to her. Teacher Casey's class of 23 included student 5.

In your district, most students attend the school in their neighborhood. Occasionally, parents request that their child attend a school other than their neighborhood school, but such requests are not always honored. Within a school, the principal assigns students to a teacher; the teachers are not allowed to request particular students.

A companion table called Year2ClassAssignments shows how students in Year 2 were assigned to the same second-grade classes used for Year 1. No students were "kept back" at the end of Year 1, so no Year 1 students reappear in a teacher's class in Year 2.

Each second-grade student in the district took standardized tests at the end of first grade, and again at the end of second grade. Scores for Year 1 students are summarized in the table Year1StudentTestScores. The first few records of the table are shown in Figure 12-3.

Year1StudentTestScores		
StudentNum ▾	FirstGradeScore ▾	SecondGradeScore ▾
1	453	456
2	386	369
3	342	354

FIGURE 12-3   Year1StudentTestScores table records

Student 1's scores on the two year-end tests were 453 and 456. Each year's test emphasizes math and reading, and each succeeding test is more comprehensive and more difficult. Thus, Student 1 did not appear to progress greatly in second grade. Student 2's second-grade score appears to represent a decline in ability. Student 3's second-grade test score shows improvement over Year 1, although the scores are not high in either year.

The Year1StudentTestScores table has 548 records, one for each Year 1 student. A companion table, Year2StudentTestScores, contains records in the same format for second-grade students in Year 2 of the model.

Your model will use two criteria for assessing teacher performance.

1. Criterion 1: The average test score is computed for a second-grade class taught by one of the sample group teachers. The average test score is also computed for all second-grade students in the district. If the sample teacher's class average is higher than the district average, the teacher is rated "good" for the year; otherwise, the teacher receives a "poor" rating for the year.

2. Criterion 2: Each student has a first-grade test score and a second-grade test score. The average test score improvement is computed for the sample teacher's class and for all second-grade students in the district. If the teacher's class has a better average improvement than the district's, the teacher is rated "good" for the year; otherwise, the teacher receives a "poor" rating for the year. For example, if a teacher's class had an average first-grade test score of 460 and an average second-grade test score of 474, the average improvement of the class is 3% (1.03 * 460 = 474). If the average district student showed a 2.5% improvement from the first-grade test to the second-grade test, the teacher would receive a "good" rating for the year.

You want to see if the two criteria produce the same ratings for a teacher in a year and if a criterion produces the same rating for a teacher two years in a row.

ASSIGNMENT 1: MAKING QUERIES IN ACCESS

In this assignment, you will design and run four queries.

Year 1 Test Scores for All District Students

Create a query whose output shows Year 1 test scores for all second-grade students in the district. Your output should look like that in Figure 12-4. The query should generate 548 output records, although only the first few are shown. Name the query Year1ScoresForDistrict.

StudentNum	LackOfInvolvement?	FirstGradeScore	SecondGradeScore
1	YES	453	456
2	YES	386	369
3	NO	342	354

FIGURE 12-4 Year 1 scores for all second-grade students in the district

You should also create a query for Year 2 test scores of all second-grade students in the district. The output format is the same.

Year 1 Test Scores for Students in Sample School

Create a query whose output shows Year 1 test scores for all second-grade students in the sample school. The query should generate 69 records for the three teachers in your model. Your output should look like that in Figure 12-5; only the first few output records are shown. Name the query Year1TestScoresForSchool.

StudentNum	LackOfInvolvement?	ClassNum	Teacher	FirstGradeScore	SecondGradeScore
1	YES	1	Smith	453	456
2	YES	1	Smith	386	369
10	YES	1	Smith	434	437
11	NO	1	Smith	428	440

FIGURE 12-5 Year 1 scores for second-grade students in sample school

You should also create a query for Year 2 test scores of all second-grade students in the sample school. The output format is the same.

When you finish the queries, save and close the **TeacherEval.accdb** file.

ASSIGNMENT 2: USING EXCEL FOR DECISION SUPPORT

In this assignment, you will import the output of the four Access queries into Excel worksheets and then develop information needed to rank the three teachers.

Importing Query Data

Open a new file in Excel and save it as **TeacherEval.xlsx**.

Import the Year1ScoresForDistrict query output into Excel. Click the Data tab and then click From Access in the Get External Data group. Specify the Access filename, the query name, and where to place the data in Excel (cell A1 is recommended).

The data will be imported into Excel as an Excel data table, which is the format you want. If cell A1 is not already selected, click it. In the Table Style Options group, select Total Row to add a Totals row to the bottom of the table. Rename the worksheet Year1ScoresForDistrict. The first few rows of your worksheet should look like Figure 12-6.

	A	B	C	D
1	StudentNum	LackOfInvolvement?	FirstGradeScore	SecondGradeScore
2	1	YES	453	456
3	2	YES	386	369
4	3	NO	342	354

FIGURE 12-6   Rows in the Year1ScoresForDistrict worksheet

Import the Year2ScoresForDistrict query output into another worksheet. Add a Totals row and name the worksheet Year2ScoresForDistrict.

Import the Year1TestScoresForSchool query output into a third worksheet. Add a Totals row and name the worksheet Year1ScoresForSchool. The first few rows of your worksheet should look like Figure 12-7.

	A	B	C	D	E	F
1	StudentNum	LackOfInvolvement?	ClassNum	Teacher	FirstGradeScore	SecondGradeScore
2	1	YES	1	Smith	453	456
3	2	YES	1	Smith	386	369
4	10	YES	1	Smith	434	437

FIGURE 12-7   Rows in the Year1ScoresForSchool worksheet

Finally, import the Year2TestScoresForSchool query output into a fourth worksheet. Add a Totals row and name the worksheet Year2ScoresForSchool.

Using Data Tables to Gather Data

You will use the data tables to gather data needed to evaluate teacher performance for the two years. You could also use pivot tables to gather the data; your instructor may require them instead. However, data tables are probably more convenient in this case.

Before gathering data, you need a worksheet in which to manually enter data that you develop. Create a new worksheet named Summary. You can use any format that lets you compare one value to another. For example, the format shown in Figure 12-8 is acceptable.

	A	B	C	D	E	F	G	H
1	Year 1 --District	**N**	**First Grade**	**SecondGrade**	**Improvement**			
2	Average--All	548						
3	LackInvolvement -- Yes							
4	LackInvolvement -- No							
5								
6	Year 2 --District	**N**	**First Grade**	**SecondGrade**	**Improvement**			
7	Average--All	548						
8	LackInvolvement -- Yes							
9	LackInvolvement -- No							
10							**Good?**	**Good?**
11	Year 1 - school	**N**	**First Grade**	**SecondGrade**	**Improvement**		**Criteria1**	**Criteria2**
12	Smith	23						
13	Jones	23						
14	Casey	23						
15	All	69						
16								
17	Year 2 - school	**N**	**First Grade**	**SecondGrade**	**Improvement**			
18	Smith	23						
19	Jones	23						
20	Casey	23						
21	All	69						
22								
23	Lacks Involvement -- Yes		**Year 1**	**Year 2**				
24	Smith							
25	Jones							
26	Casey							

FIGURE 12-8 Possible summary worksheet format

Most data for the summary worksheet will come from the data tables. Data in the Improvement column would be determined by an Excel formula in which one cell value is divided by another. The Criteria cells should include an explanatory comment; to enter one, right-click in the cell and click Insert Comment. The bottom section, "Lacks Involvement—Yes," records counts of students whose parents lacked involvement in each teacher's classes during the two years. For example, Smith might have had 8 uninvolved parents in Year 1 and 12 in Year 2.

You should now gather data for the sample teachers' classes and for the district, using the four data tables. As data is gathered, enter the values manually into your summary sheet. Tutorial E explains how to use data tables. Keep the following points in mind:

- Click the arrow in the Totals row to select summary values for a column's data, such as averages or counts.
- Column headings also have arrows. You can stratify data in a column by clicking a value on or off. For example, suppose that your worksheet showed people's heights and gender. You could click to hide information for males, leaving just female data. Then you could use the Totals row to determine the average height for females. To restore the male data, you would click the heading arrow and then click Select All.

Fill in the Criteria cells after inspecting the data. Then consider the following eight questions:

1. What is the impact of parental involvement? Is it true that children of involved parents do better academically in this school district?
2. Using Criterion 1, are any teachers rated as good in one year but poor in the other? If so, does it seem reasonable that a teacher's ability changes from year to year? How can these rating changes be explained?
3. Using Criterion 1, does it appear that any teachers are classified incorrectly? In other words, are any teachers considered good when they actually are poor, or vice versa?
4. Is Criterion 1 a reasonable way to assess teacher ability? Why or why not?
5. Using Criterion 2, are any teachers rated as good in one year but poor in the other? If so, does it seem reasonable that a teacher's ability changes from year to year? How can these rating changes be explained?

6. Using Criterion 2, does it appear that any teachers are classified incorrectly? In other words, are any teachers considered good when they actually are poor, or vice versa?

7. Is Criterion 2 a reasonable way to assess teacher ability? Why or why not?

8. Based on the data, how would you describe the abilities of each teacher?

ASSIGNMENT 3: DOCUMENTING FINDINGS IN A MEMORANDUM

In this assignment, you write a memorandum in Microsoft Word that documents your findings. Your memo needs tables that show teacher ratings by the two criteria for each year. Tutorial E describes how to create a table in Microsoft Word. The format of the tables is shown in Figure 12-9; you will also need a table for Year 2.

Teacher	Year 1 Rating by Criterion 1	Year 1 Rating by Criterion 2
Smith		
Jones		
Casey		

FIGURE 12-9   Format of table to include in memo

In your memo, observe the following requirements:

- Your memo should have proper headings, such as Date, To, From, and Subject. You can address the memo to the district superintendent. Set up the memo as discussed in Tutorial E.
- Briefly outline the situation. However, you need not provide much background—you can assume that readers are generally familiar with your task.
- In the body of the memo, briefly describe the model you created. Be sure to state the two accountability criteria. List the ratings given to the three teachers, and refer to the tables that summarize the ratings.
- List the answers to the eight questions. Then state your opinion of measuring teacher accountability by standardized test scores. If the method has problems, how do you think the problems can be minimized?

DELIVERABLES

Assemble the following deliverables for your instructor:

1. Printout of your memo
2. Spreadsheet printouts, if required by your instructor
3. Electronic media such as flash drive or CD, which should include your Word file, Access file, and Excel file

Staple the printouts together with the memo on top. If you have more than one .xlsx file or .accdb file on your flash drive or CD, write your instructor a note that identifies the files in this assignment.

PART 6

ADVANCED SKILLS USING EXCEL

TUTORIAL E
Guidance for Excel Cases, 241

GUIDANCE FOR EXCEL CASES

The Microsoft Excel cases in this book require the student to write a memorandum that includes a table. Guidelines for preparing a memo in Microsoft Word and instructions for entering a table in a Word document are provided to begin this tutorial. Also, some of the cases in this book require the use of advanced Excel techniques. Those techniques are explained in this tutorial rather than in the cases themselves:

- Using data tables
- Using pivot tables
- Using built-in functions

You can refer to Sheet 1 of TutEData.xlsx when reading about data tables. Refer to Sheet 2 when reading about pivot tables.

PREPARING A MEMORANDUM IN WORD

A business memo should include proper headings, such as TO, FROM, DATE, and SUBJECT. If you want to use a Word memo template, follow these steps:

1. In Word, click File.
2. Click New.
3. Click the Memos button in the Office.com or Microsoft Office Online Templates section.
4. Double-click the Contemporary design memo.

The first time you do this, you may need to click Download to install the template.

ENTERING A TABLE INTO A WORD DOCUMENT

Enter a table into a Word document using the following procedure:

1. Click the cursor where you want the table to appear in the document.
2. In the Insert group, click the Table drop-down menu.
3. Click Insert Table.
4. Choose the number of rows and columns.
5. Click OK.

DATA TABLES

An Excel data table is a contiguous range of data that has been designated as a table. Once you make this designation, the table gains certain properties that are useful for data analysis. (Note that in some previous versions of Excel, data tables were called *data lists*.) Suppose you have a list of runners who have completed a race, as shown in Figure E-1.

	A	B	C	D	E	F
1	RUNNER#	LAST	FIRST	AGE	GENDER	TIME (MIN)
2	100	HARRIS	JANE	O	F	70
3	101	HILL	GLENN	Y	M	70
4	102	GARCIA	PEDRO	M	M	85
5	103	HILBERT	DORIS	M	F	90
6	104	DOAKS	SALLY	Y	F	94
7	105	JONES	SUE	Y	F	95
8	106	SMITH	PETE	M	M	100
9	107	DOE	JANE	O	F	100
10	108	BRADY	PETE	O	M	100
11	109	BRADY	JOE	O	M	120
12	110	HEEBER	SALLY	M	F	125
13	111	DOLTZ	HAL	O	M	130
14	112	PEEBLES	AL	Y	M	63

FIGURE E-1 Data table example

To turn the information into a data table, highlight the data range, including headings, and click the Insert tab. Then click Table in the Tables group. The Create Table window appears, as shown in Figure E-2.

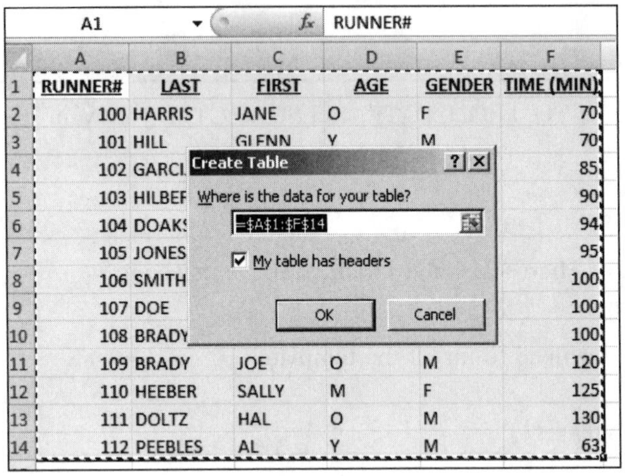

FIGURE E-2 Create Table window

When you click OK, the data range appears as a table. On the Design tab, click Total Row to add a totals row to the data table. You also can select a light style in the Table Styles list to get rid of the contrasting color in the table's rows. Figure E-3 shows the results.

	A	B	C	D	E	F
1	RUNNER#	LAST	FIRST	AGE	GENDER	TIME (MIN)
2	100	HARRIS	JANE	O	F	70
3	101	HILL	GLENN	Y	M	70
4	102	GARCIA	PEDRO	M	M	85
5	103	HILBERT	DORIS	M	F	90
6	104	DOAKS	SALLY	Y	F	94
7	105	JONES	SUE	Y	F	95
8	106	SMITH	PETE	M	M	100
9	107	DOE	JANE	O	F	100
10	108	BRADY	PETE	O	M	100
11	109	BRADY	JOE	O	M	120
12	110	HEEBER	SALLY	M	F	125
13	111	DOLTZ	HAL	O	M	130
14	112	PEEBLES	AL	Y	M	63
15	Total					1242

FIGURE E-3 Data table example

The headings have acquired drop-down menu tabs, as you can see in Figure E-3.

You can sort the data table records by any field. Perhaps you want to sort by times. If so, click the drop-down menu in the TIME (MIN) heading, select Sort, and then click Smallest to Largest. You get the results shown in Figure E-4.

	A	B	C	D	E	F
1	RUNNER ▾	LAST ▾	FIRST ▾	AGE ▾	GENDE ▾	TIME (MIN ▾1
2	112	PEEBLES	AL	Y	M	63
3	100	HARRIS	JANE	O	F	70
4	101	HILL	GLENN	Y	M	70
5	102	GARCIA	PEDRO	M	M	85
6	103	HILBERT	DORIS	M	F	90
7	104	DOAKS	SALLY	Y	F	94
8	105	JONES	SUE	Y	F	95
9	106	SMITH	PETE	M	M	100
10	107	DOE	JANE	O	F	100
11	108	BRADY	PETE	O	M	100
12	109	BRADY	JOE	O	M	120
13	110	HEEBER	SALLY	M	F	125
14	111	DOLTZ	HAL	O	M	130
15	Total					1242

FIGURE E-4 Sorting list by drop-down menu

You can see that Peebles had the best time and Doltz had the worst time. You also can sort from Largest to Smallest.

In addition, you can sort by more than one criterion. Assume that you want to sort first by gender and then by time (within gender). You first sort from Smallest to Largest in Gender. Then you again click the Gender drop-down tab, click Sort By Color, and then click Custom Sort. In the Sort window that appears, click Add Level and choose Time as the next criterion. See Figure E-5.

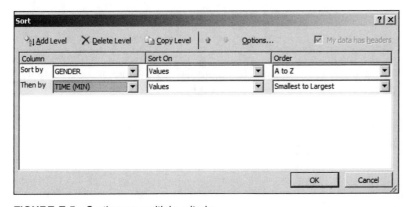

FIGURE E-5 Sorting on multiple criteria

Click OK to get the results shown in Figure E-6.

	A	B	C	D	E	F
1	RUNNER ▾	LAST ▾	FIRST ▾	AGE ▾	GENDE ▾↑	TIME (MIN ▾↑
2	100	HARRIS	JANE	O	F	70
3	103	HILBERT	DORIS	M	F	90
4	104	DOAKS	SALLY	Y	F	94
5	105	JONES	SUE	Y	F	95
6	107	DOE	JANE	O	F	100
7	110	HEEBER	SALLY	M	F	125
8	112	PEEBLES	AL	Y	M	63
9	101	HILL	GLENN	Y	M	70
10	102	GARCIA	PEDRO	M	M	85
11	106	SMITH	PETE	M	M	100
12	108	BRADY	PETE	O	M	100
13	109	BRADY	JOE	O	M	120
14	111	DOLTZ	HAL	O	M	130
15	Total					1242

FIGURE E-6   Sorting by gender and time (within gender)

You can see that Harris had the best female time and that Peebles had the best male time.

Perhaps you want to see the top *n* listings for some attribute; for example, you may want to see the top five runners' times. Select the Time column's drop-down menu, and select Number Filters. From the menu that appears, select Top 10. The Top 10 AutoFilter window appears, as shown in Figure E-7.

FIGURE E-7   Top 10 AutoFilter window

This window lets you specify the number of values you want. In Figure E-7, five values were specified. Click OK to get the results shown in Figure E-8.

	A	B	C	D	E	F
1	RUNNER ▾	LAST ▾	FIRST ▾	AGE ▾	GENDE ▾↑	TIME (MIN ▾
6	107	DOE	JANE	O	F	100
7	110	HEEBER	SALLY	M	F	125
11	106	SMITH	PETE	M	M	100
12	108	BRADY	PETE	O	M	100
13	109	BRADY	JOE	O	M	120
14	111	DOLTZ	HAL	O	M	130
15	Total					675

FIGURE E-8   Top 5 times

The output contains more than five data records because there are ties at 100 minutes. If you want to see all of the records again, click the Time drop-down menu and click Clear Filter. The full table of data reappears, as shown in Figure E-9.

	A	B	C	D	E	F
1	**RUNNER** ▾	**LAST** ▾	**FIRST** ▾	**AGE** ▾	**GENDE** ▾↑	**TIME (MIN** ▾↑
2	100	HARRIS	JANE	O	F	70
3	103	HILBERT	DORIS	M	F	90
4	104	DOAKS	SALLY	Y	F	94
5	105	JONES	SUE	Y	F	95
6	107	DOE	JANE	O	F	100
7	110	HEEBER	SALLY	M	F	125
8	112	PEEBLES	AL	Y	M	63
9	101	HILL	GLENN	Y	M	70
10	102	GARCIA	PEDRO	M	M	85
11	106	SMITH	PETE	M	M	100
12	108	BRADY	PETE	O	M	100
13	109	BRADY	JOE	O	M	120
14	111	DOLTZ	HAL	O	M	130
15	Total					1242

FIGURE E-9   Restoring all data to screen

Each of the cells in the Total row has a drop-down menu. The menu choices are statistical operations that you can perform on the totals—for example, you can take a sum, take an average, take a minimum or maximum, count the number of records, and so on. Assume that the Time drop-down menu was selected, as shown in Figure E-10. Note that the Sum operator is highlighted by default.

	A	B	C	D	E	F
1	**RUNNER** ▾	**LAST** ▾	**FIRST** ▾	**AGE** ▾	**GENDE** ▾↑	**TIME (MIN** ▾↑
2	100	HARRIS	JANE	O	F	70
3	103	HILBERT	DORIS	M	F	90
4	104	DOAKS	SALLY	Y	F	94
5	105	JONES	SUE	Y	F	95
6	107	DOE	JANE	O	F	100
7	110	HEEBER	SALLY	M	F	125
8	112	PEEBLES	AL	Y	M	63
9	101	HILL	GLENN	Y	M	70
10	102	GARCIA	PEDRO	M	M	85
11	106	SMITH	PETE	M	M	100
12	108	BRADY	PETE	O	M	100
13	109	BRADY	JOE	O	M	120
14	111	DOLTZ	HAL	O	M	130
15	Total					1242 ▾
16						None
17						Average
18						Count
						Count Numbers
19						Max
20						Min
21						Sum
						StdDev
22						Var
						More Functions...

FIGURE E-10   Selecting Time drop-down menu in Total row

By changing from Sum to Average, you find that the average time for all runners was 95.5 minutes, as shown in Figure E-11.

	A	B	C	D	E	F
1	RUNNER ▾	LAST ▾	FIRST ▾	AGE ▾	GENDE ▾1	TIME (MIN ▾1
2	100	HARRIS	JANE	O	F	70
3	103	HILBERT	DORIS	M	F	90
4	104	DOAKS	SALLY	Y	F	94
5	105	JONES	SUE	Y	F	95
6	107	DOE	JANE	O	F	100
7	110	HEEBER	SALLY	M	F	125
8	112	PEEBLES	AL	Y	M	63
9	101	HILL	GLENN	Y	M	70
10	102	GARCIA	PEDRO	M	M	85
11	106	SMITH	PETE	M	M	100
12	108	BRADY	PETE	O	M	100
13	109	BRADY	JOE	O	M	120
14	111	DOLTZ	HAL	O	M	130
15	Total					95.53846154 ▾

FIGURE E-11   Average running time shown in Total row

PIVOT TABLES

Suppose you have data for a company's sales transactions by month, by salesperson, and by amount for each product type. You would like to display each salesperson's total sales by type of product sold and by month. You can use a pivot table in Excel to tabulate that summary data. A pivot table is built around one or more dimensions and thus can summarize large amounts of data. Figure E-12 shows total sales cross-tabulated by salesperson and by month.

	A	B	C	D	E
1	**Name**	**Product**	**January**	**February**	**March**
2	Jones	Product1	30,000	35,000	40,000
3	Jones	Product2	33,000	34,000	45,000
4	Jones	Product3	24,000	30,000	42,000
5	Smith	Product1	40,000	38,000	36,000
6	Smith	Product2	41,000	37,000	38,000
7	Smith	Product3	39,000	50,000	33,000
8	Bonds	Product1	25,000	26,000	25,000
9	Bonds	Product2	22,000	25,000	24,000
10	Bonds	Product3	19,000	20,000	19,000
11	Ruth	Product1	44,000	42,000	33,000
12	Ruth	Product2	45,000	40,000	30,000
13	Ruth	Product3	50,000	52,000	35,000

FIGURE E-12   Excel spreadsheet data

You can create pivot tables and many other kinds of tables with the Excel PivotTable tool. To create a pivot table from the data in Figure E-12, follow these steps:

1.   Starting in the spreadsheet in Figure E-12, click a cell in the data range, and then click the Insert tab. In the Tables group, choose PivotTable. You see the screen shown in Figure E-13.

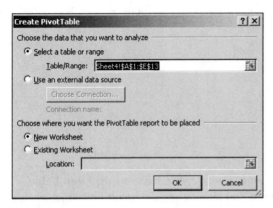

FIGURE E-13 Creating a pivot table

2. Make sure New Worksheet is checked under "Choose where you want the PivotTable report to be placed." Click OK. The screen shown in Figure E-14 appears. If it does not, right-click in a cell in the pivot table area, and click PivotTable Options from the menu that appears. Click the Display tab and then check the Classic PivotTable layout.

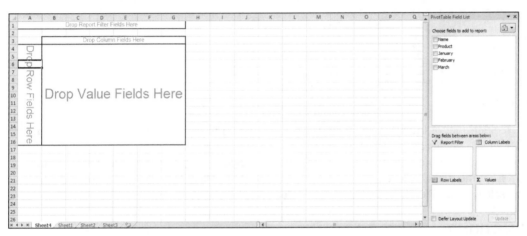

FIGURE E-14 PivotTable design screen

The data range's column headings are shown in the PivotTable Field List on the right side of the screen. From there, you can click and drag column headings into the Row, Column, and Value areas that appear in the spreadsheet.

3. If you want to see the total sales by product for each salesperson, drag the Name field to the Drop Column Fields Here area in the spreadsheet. You should see the result shown in Figure E-15.

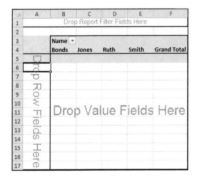

FIGURE E-15 Column fields

4. Next, drag the Product field to the Drop Row Fields Here area. You should see the result shown in Figure E-16.

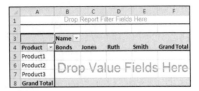

FIGURE E-16   Row fields

5. Finally, drag the month fields (January, February, and March) individually to the Drop Value Fields Here area to produce the finalized pivot table. You should see the result shown in Figure E-17.

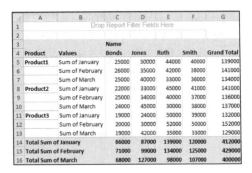

FIGURE E-17   Data items

By default, Excel adds all of the sales for each salesperson by month for each product. At the bottom of the pivot table, Excel also shows the total sales for each month for all products.

Refer back to Figure E-14 and note the four small panes in the lower-right corner. The Values pane lets you easily change from the default Sum operator to another one (Min, Max, Average, and so on). Click the drop-down arrow, select Value Fields Setting, and then select the desired operator.

BUILT-IN FUNCTIONS

You might need to use some of the following functions when solving the Excel cases elsewhere in this text:

- MIN, MAX, AVERAGE, COUNTIF, ROUND, ROUNDUP, and RANDBETWEEN

The syntax of these functions is discussed in this section. The following examples are based on the runner data shown in Figure E-18.

	RUNNER#	LAST	FIRST	AGE	GENDER	HEIGHT (IN)	TIME (MIN)
1							
2	100	HARRIS	JANE	O	F	60	70
3	101	HILL	GLENN	Y	M	65	70
4	102	GARCIA	PEDRO	M	M	76	85
5	103	HILBERT	DORIS	M	F	64	90
6	104	DOAKS	SALLY	Y	F	62	94
7	105	JONES	SUE	Y	F	64	95
8	106	SMITH	PETE	M	M	73	100
9	107	DOE	JANE	O	F	66	100
10	108	BRADY	PETE	O	M	73	100
11	109	BRADY	JOE	O	M	71	120
12	110	HEEBER	SALLY	M	F	59	125
13	111	DOLTZ	HAL	O	M	76	130
14	112	PEEBLES	AL	Y	M	76	63

FIGURE E-18   Runner data used to illustrate built-in functions

The data is the same as that shown in Figure E-1, except that Figure E-18 includes a column for the runners' height in inches.

MIN and MAX Functions

The MIN function determines the smallest value in a range of data. The MAX function returns the largest. Say that we want to know the fastest time for all runners, which would be the minimum time in column G. The MIN function computes the smallest value in a set of values. The set of values could be a data range, or it could be a series of cell addresses separated by commas. The syntax of the MIN function is as follows:

- MIN(set of data)

To show the minimum time in cell C16, you would enter the formula shown in the formula bar in Figure E-19:

	C16	▼	f_x	=MIN(G2:G14)		
	A	B	C	D	E	F	G
1	RUNNER#	LAST	FIRST	AGE	GENDER	HEIGHT	TIME (MIN)
2	100	HARRIS	JANE	O	F	60	70
3	101	HILL	GLENN	Y	M	65	70
4	102	GARCIA	PEDRO	M	M	76	85
5	103	HILBERT	DORIS	M	F	64	90
6	104	DOAKS	SALLY	Y	F	62	94
7	105	JONES	SUE	Y	F	64	95
8	106	SMITH	PETE	M	M	73	100
9	107	DOE	JANE	O	F	66	100
10	108	BRADY	PETE	O	M	73	100
11	109	BRADY	JOE	O	M	71	120
12	110	HEEBER	SALLY	M	F	59	125
13	111	DOLTZ	HAL	O	M	76	130
14	112	PEEBLES	AL	Y	M	76	63
15							
16	MINIMUM TIME:		63				

FIGURE E-19   MIN function in cell C16

(Assume that you typed the label "MINIMUM TIME:" into cell A16.) You can see that the fastest time is 63 minutes.

To see the slowest time in cell G16, use the MAX function, whose syntax parallels that of the MIN function, except that the largest value in the set is determined. See Figure E-20.

	G16	▼	f_x	=MAX(G2:G14)		
	A	B	C	D	E	F	G
1	RUNNER#	LAST	FIRST	AGE	GENDER	HEIGHT	TIME (MIN)
2	100	HARRIS	JANE	O	F	60	70
3	101	HILL	GLENN	Y	M	65	70
4	102	GARCIA	PEDRO	M	M	76	85
5	103	HILBERT	DORIS	M	F	64	90
6	104	DOAKS	SALLY	Y	F	62	94
7	105	JONES	SUE	Y	F	64	95
8	106	SMITH	PETE	M	M	73	100
9	107	DOE	JANE	O	F	66	100
10	108	BRADY	PETE	O	M	73	100
11	109	BRADY	JOE	O	M	71	120
12	110	HEEBER	SALLY	M	F	59	125
13	111	DOLTZ	HAL	O	M	76	130
14	112	PEEBLES	AL	Y	M	76	63
15							
16	MINIMUM TIME:		63		MAXIMUM TIME:		130

FIGURE E-20   MAX function in cell G16

AVERAGE, ROUND, and ROUNDUP Functions

The AVERAGE function computes the average of a set of values. Figure E-21 shows the use of the AVERAGE function in cell C17:

	C17		▼	f_x	=AVERAGE(G2:G14)		
	A	B	C	D	E	F	G
1	RUNNER#	LAST	FIRST	AGE	GENDER	HEIGHT	TIME (MIN)
2	100	HARRIS	JANE	O	F	60	70
3	101	HILL	GLENN	Y	M	65	70
4	102	GARCIA	PEDRO	M	M	76	85
5	103	HILBERT	DORIS	M	F	64	90
6	104	DOAKS	SALLY	Y	F	62	94
7	105	JONES	SUE	Y	F	64	95
8	106	SMITH	PETE	M	M	73	100
9	107	DOE	JANE	O	F	66	100
10	108	BRADY	PETE	O	M	73	100
11	109	BRADY	JOE	O	M	71	120
12	110	HEEBER	SALLY	M	F	59	125
13	111	DOLTZ	HAL	O	M	76	130
14	112	PEEBLES	AL	Y	M	76	63
15							
16	MINIMUM TIME:		63		MAXIMUM TIME:		130
17	AVERAGE TIME:		95.53846				

FIGURE E-21   AVERAGE function in cell C17

Notice that the value shown is a real number with many digits. What if you wanted to have the value rounded to a certain number of digits? Of course, you could format the output cell, but doing that changes only what is shown on the screen. You want the cell's contents actually to *be* the rounded number. Therefore, you need to use the ROUND function. Its syntax is:

- ROUND(number, number of digits)

Figure E-22 shows the rounded average time (with two decimal places) in cell G17.

	G17		▼	f_x	=ROUND(C17,2)		
	A	B	C	D	E	F	G
1	RUNNER#	LAST	FIRST	AGE	GENDER	HEIGHT	TIME (MIN)
2	100	HARRIS	JANE	O	F	60	70
3	101	HILL	GLENN	Y	M	65	70
4	102	GARCIA	PEDRO	M	M	76	85
5	103	HILBERT	DORIS	M	F	64	90
6	104	DOAKS	SALLY	Y	F	62	94
7	105	JONES	SUE	Y	F	64	95
8	106	SMITH	PETE	M	M	73	100
9	107	DOE	JANE	O	F	66	100
10	108	BRADY	PETE	O	M	73	100
11	109	BRADY	JOE	O	M	71	120
12	110	HEEBER	SALLY	M	F	59	125
13	111	DOLTZ	HAL	O	M	76	130
14	112	PEEBLES	AL	Y	M	76	63
15							
16	MINIMUM TIME:		63		MAXIMUM TIME:		130
17	AVERAGE TIME:		95.53846		ROUNDED AVERAGE		95.54

FIGURE E-22   ROUND function used in cell G17

To achieve this output, cell C17 was used as the value to be rounded. Recall from Figure E-21 that cell C17 had the formula =AVERAGE(G2:G14). The following ROUND formula would produce the same output in cell G17: =ROUND(AVERAGE(G2:G14),2). In this case, Excel evaluates the formula "inside out." First, the AVERAGE function is evaluated, yielding the average with many digits. That value is then input to the ROUND function and rounded to two decimal places.

The ROUNDUP function works much like the ROUND function. ROUNDUP's output is always rounded up to the next value. For example, the value 4 would appear in a cell that contained the following formula: =ROUNDUP(3.12,0). In Figure E-22, if the formula in cell G17 had been =ROUNDUP(AVERAGE(G2:G14),2), the value 96 would have been the result. In other words, 95.54 rounded up with no decimal places becomes 96.

COUNTIF Function

The COUNTIF function counts the number of values in a range that meet a specified condition. The syntax is:

- COUNTIF(range of data, condition)

The condition is a logical expression such as "=1", ">6", or "=F". The condition is shown with quotation marks, even if a number is involved.

Assume that you want to see the number of female runners in cell C18. Figure E-23 shows the formula used.

	C18	▾	f_x	=COUNTIF(E2:E14,"F")		
	A	B	C	D	E	F	G
1	RUNNER#	LAST	FIRST	AGE	GENDER	HEIGHT	TIME (MIN)
2	100	HARRIS	JANE	O	F	60	70
3	101	HILL	GLENN	Y	M	65	70
4	102	GARCIA	PEDRO	M	M	76	85
5	103	HILBERT	DORIS	M	F	64	90
6	104	DOAKS	SALLY	Y	F	62	94
7	105	JONES	SUE	Y	F	64	95
8	106	SMITH	PETE	M	M	73	100
9	107	DOE	JANE	O	F	66	100
10	108	BRADY	PETE	O	M	73	100
11	109	BRADY	JOE	O	M	71	120
12	110	HEEBER	SALLY	M	F	59	125
13	111	DOLTZ	HAL	O	M	76	130
14	112	PEEBLES	AL	Y	M	76	63
15							
16	MINIMUM TIME:		63		MAXIMUM TIME:		130
17	AVERAGE TIME:		95.53846		ROUNDED AVERAGE:		95.54
18	NUMBER OF FEMALES:		6				

FIGURE E-23   COUNTIF function used in cell C18

The logic of the formula is: Count the number of times that "F" appears in the data range E2:E14.

As another example of using COUNTIF, assume that column H shows the rounded ratio of each runner's height in inches to the runner's time in minutes (see Figure E-24).

	H2	▾		f_x	=ROUND(G2/F2,2)			
	A	B	C	D	E	F	G	H
1	RUNNER#	LAST	FIRST	AGE	GENDER	HEIGHT	TIME (MIN)	RATIO
2	100	HARRIS	JANE	O	F	60	70	1.17
3	101	HILL	GLENN	Y	M	65	70	1.08
4	102	GARCIA	PEDRO	M	M	76	85	1.12
5	103	HILBERT	DORIS	M	F	64	90	1.41
6	104	DOAKS	SALLY	Y	F	62	94	1.52
7	105	JONES	SUE	Y	F	64	95	1.48
8	106	SMITH	PETE	M	M	73	100	1.37
9	107	DOE	JANE	O	F	66	100	1.52
10	108	BRADY	PETE	O	M	73	100	1.37
11	109	BRADY	JOE	O	M	71	120	1.69
12	110	HEEBER	SALLY	M	F	59	125	2.12
13	111	DOLTZ	HAL	O	M	76	130	1.71
14	112	PEEBLES	AL	Y	M	76	63	0.83
15								
16	MINIMUM TIME:		63		MAXIMUM TIME:		130	
17	AVERAGE TIME:		95.53846		ROUNDED AVERAGE:		95.54	
18	NUMBER OF FEMALES:		6					

FIGURE E-24   Ratio of height to time in column H

Assume that all runners whose height in inches is less than their time in minutes will get an award. How many awards are needed? If the ratio is less than 1, an award is warranted. The COUNTIF function in cell G18 computes a count of ratios less than 1, as shown in Figure E-25.

	G18	▾		f_x	=COUNTIF(H2:H14,"<1")			
	A	B	C	D	E	F	G	H
1	RUNNER#	LAST	FIRST	AGE	GENDER	HEIGHT	TIME (MIN)	RATIO
2	100	HARRIS	JANE	O	F	60	70	1.17
3	101	HILL	GLENN	Y	M	65	70	1.08
4	102	GARCIA	PEDRO	M	M	76	85	1.12
5	103	HILBERT	DORIS	M	F	64	90	1.41
6	104	DOAKS	SALLY	Y	F	62	94	1.52
7	105	JONES	SUE	Y	F	64	95	1.48
8	106	SMITH	PETE	M	M	73	100	1.37
9	107	DOE	JANE	O	F	66	100	1.52
10	108	BRADY	PETE	O	M	73	100	1.37
11	109	BRADY	JOE	O	M	71	120	1.69
12	110	HEEBER	SALLY	M	F	59	125	2.12
13	111	DOLTZ	HAL	O	M	76	130	1.71
14	112	PEEBLES	AL	Y	M	76	63	0.83
15								
16	MINIMUM TIME:		63		MAXIMUM TIME:		130	
17	AVERAGE TIME:		95.53846		ROUNDED AVERAGE:		95.54	
18	NUMBER OF FEMALES:		6		RATIOS<1:		1	

FIGURE E-25   COUNTIF function used in cell G18

RANDBETWEEN Function

If you wanted a cell to contain a randomly generated integer in the range from 1 to 9, you would use the formula =RANDBETWEEN(1,9). Any value between 1 and 9 inclusive would be output by the formula. An example is shown in Figure E-26.

A2		f_x	=RANDBETWEEN(1,9)		
	A	B	C	D	E
1	Position				
2	9				

FIGURE E-26   RANDBETWEEN function used in cell A2

Assume that you copied and pasted the formula to generate a column of 100 numbers between 1 and 9. Every time a value was changed in the spreadsheet, Excel would recalculate the 100 RANDBETWEEN formulas to change the 100 random values. Therefore, you might want to settle on the random values once they are generated. To do this, copy the 100 values, click Paste Special, and then click Values to put the values in the same range. The contents of the cells will change from formulas to literal values.

PART 7

PRESENTATION SKILLS

GIVING AN ORAL PRESENTATION

Giving an oral presentation in class lets you practice the presentation skills you will need in the workplace. The presentations you create for the cases in this textbook will be similar to professional business presentations. You will be expected to present objective, technical results to your organization's stakeholders, and you will have to support your presentation with visual aids commonly used in the business world. During your presentation, your instructor might assign your classmates to role-play an audience of business managers, bankers, or employees. They might also provide feedback on your presentation.

Follow these four steps to create an effective presentation:

1. Plan your presentation.
2. Draft your presentation.
3. Create graphics and other visual aids.
4. Practice delivering your presentation.

PLANNING YOUR PRESENTATION

When planning an oral presentation, you need to know your time limits, establish your purpose, analyze your audience, and gather information. This section explores each of these elements.

Knowing Your Time Limits

You need to consider your time limits on two levels. First, consider how much time you will have to deliver your presentation. For example, what are the key points in your material that can be covered in 10 minutes? The element of time is the primary constraint of any presentation. It limits the breadth and depth of your talk, and the number of visual aids that you can use. Second, consider how much time you will need for the process of preparing your presentation—drafting your presentation, creating graphics, and practicing your delivery.

Establishing Your Purpose

After considering your time limits, you must define your purpose: what you need to say and to whom you will say it. For the Access cases in this book, your purpose will be to inform and explain. For instance, a business's owners, managers, and employees may need to know how the company's database is organized and how they can use it to fill in forms and create reports. In contrast, for the Excel cases, your purpose will be to recommend a course of action based on the results of your business model. You will make the recommendations to business owners, managers, and bankers based on the results of inputting and running various scenarios.

Analyzing Your Audience

Once you have established the purpose of your presentation, you should analyze your audience. Ask yourself: What does my audience already know about the subject? What do the audience members want to know? What do they need to know? Do they have any biases or personal agendas that I should consider? What level of technical detail is best suited to their level of knowledge and interest?

In some Access cases, you will make a presentation to an audience that might not be familiar with Access or with databases in general. In other cases, you might be giving your presentation to a business owner who started to work on a database but was not able to finish it. Tailor the presentation to suit your audience.

For the Excel cases, you are most often interpreting results for an audience of bankers or business managers. In those instances, the audience will not need to know the detailed technical aspects of how you generated your results. But what if your audience consists of engineers or scientists? They will certainly be more interested in the structure and rationale of your decision models. Regardless of the audience, your listeners need to know what assumptions you made prior to developing your spreadsheets because those assumptions might affect their opinion of your results.

Gathering Information

Because you will have just completed a case as you begin preparing your oral presentation, you will already have the basic information you need. For the Access cases, you should review the main points of the case and your goals. Make sure you include all of the points you think are important for the audience to understand. In addition, you might want to go beyond the requirements and explain additional ways in which the database could be used to benefit the organization, now or in the future.

For the Excel cases, you can refer to the tutorials for assistance in interpreting the results from your spreadsheet analysis. For some cases, you might want to use the Internet or the library to research business trends or background information that can support your presentation.

DRAFTING YOUR REPORT AND PRESENTATION

When you have completed the planning stage, you are ready to begin drafting the presentation. At this point, you might be tempted to write your presentation and then memorize it word for word. Even if you could memorize your presentation verbatim, however, your delivery would sound unnatural because people use a simpler vocabulary and shorter sentences when they speak than when they write. For example, read the previous paragraph out loud as if you were presenting it to an audience.

In many business situations, you will be required both to submit a written report of your work and give a PowerPoint presentation. First, write your report, and then design your PowerPoint slides as a "brief" of that report to discuss its main points. When drafting your report and the accompanying PowerPoint slides, follow this sequence:

1. Write the main body of your report.
2. Write the introduction to your report.
3. Write the conclusion to your report.
4. Prepare your presentation (the PowerPoint slides) using your report's main points.

Writing the Main Body

When you draft your report, write the body first. If you try to write the opening paragraph first, you might spend an inordinate amount of time attempting to craft your words perfectly, only to revise the introduction after you write the body of the report.

Keeping Your Audience in Mind

To write the main body, review your purpose and your audience profile. What are the main points you need to make? What are your audience's needs, interests, and technical expertise? It is important to include some technical details in your report and presentation, but keep in mind the technical expertise of your audience.

Remember that the people reading your report or listening to your presentation have their own agendas—put yourself in their places and ask, "What do I need to get out of this presentation?" For example, in the Access cases, an employee might want to know how to enter information on a form, but the business owner might be more interested in generating queries and reports. You need to address their different needs in your presentation. For example, you might say, "And now, let's look at how data entry associates can input data into this form."

Similarly, in the Excel cases, your audience will consist of business owners, managers, bankers, and perhaps some technical professionals. The owners and managers will be concerned with profitability, growth, and customer service. In contrast, the bankers' main concern will be repayment of a loan. Technical professionals will be more concerned with how well your decision model is designed, along with the credibility of the results. You need to address the interests of each group.

Using Transitions and Repetition in your Presentation

During your presentation, remember that the audience is not reading the text of your report, so you need to include transitions to compensate. Words such as *next, first, second,* and *finally* will help the audience follow the sequence of your ideas. Words such as *however, in contrast, on the other hand*, and *similarly* will help the audience follow shifts in thought. You can use your voice to convey emphasis.

Also consider using hand gestures to emphasize what you say. For instance, if you list three items, you can use your fingers to tick off each item as you discuss it. Similarly, if you state that profits will be flat, you can make a level motion with your hand for emphasis.

You may be speaking behind a podium or standing beside a projection screen, or both. If you feel uncomfortable standing in one place and you can walk without blocking the audience's view of the screen, feel free to move around. You can emphasize a transition by changing your position. If you tend to fidget, shift, or rock from one foot to the other, try to anchor yourself. A favorite technique of some speakers is to come from behind the podium and place one hand on it while speaking. They get the anchoring effect of the podium while removing the barrier it places between them and the audience. Use the stance or technique that makes you feel most comfortable, as long as your posture or actions do not distract the audience.

As you draft your presentation, repeat key points to emphasize them. For example, suppose your main point is that outsourcing labor will provide the greatest gains in net income. Begin by previewing that concept, and state that you will demonstrate how outsourcing labor will yield the biggest profits. Then provide statistics that support your claim, and show visual aids that graphically illustrate your point. Summarize by repeating your point: "As you can see, outsourcing labor does yield the biggest profits."

Relying on Graphics to Support Your Talk

As you write the main body, think of how to integrate graphics into your presentation. Do not waste words with a long description if a graphic can bring instant comprehension. For instance, instead of describing how information from a query can be turned into a report, show the query and a completed report. Figures F-1 and F-2 illustrate an Access query and the resulting report.

Order Query 1					
Customer Name	City	Product Name	Qty	Price per Unit	Total
Applewood Restaurant	Martinsburg	Frozen Alligator on a Stick	20	$27.99	$559.80
Applewood Restaurant	Martinsburg	Nogales Chipotle Sauce	15	$11.49	$172.35
Applewood Restaurant	Martinsburg	Mom's Deep Dish Apple Pie	12	$12.49	$149.88
Fresh Catch Fishery	Salem	Brumley's Seafood Cocktail Sauce	24	$4.79	$114.96
Fresh Catch Fishery	Salem	NY Smoked Salmon	21	$21.99	$461.79
Fresh Catch Fishery	Salem	Mama Mia's Tiramisu	15	$17.99	$269.85
Jimmy's Crab House	Elkton	Frozen Alligator on a Stick	12	$27.99	$335.88
Jimmy's Crab House	Elkton	Brumley's Seafood Cocktail Sauce	24	$4.79	$114.96
Jimmy's Crab House	Elkton	Mama Mia's Tiramisu	18	$17.99	$323.82
Jimmy's Crab House	Elkton	Mom's Deep Dish Apple Pie	36	$12.49	$449.64
*					

FIGURE F-1 Access query

Customer Name	City	Product Name	Qty	Price per Unit	Total
Applewood Restaurant	Martinsburg	Frozen Alligator on a Stick	20	$27.99	$559.80
Applewood Restaurant	Martinsburg	Nogales Chipotle Sauce	15	$11.49	$172.35
Applewood Restaurant	Martinsburg	Mom's Deep Dish Apple Pie	12	$12.49	$149.88
Fresh Catch Fishery	Salem	Brumley's Seafood Cocktail Sauce	24	$4.79	$114.96
Fresh Catch Fishery	Salem	NY Smoked Salmon	21	$21.99	$461.79
Fresh Catch Fishery	Salem	Mama Mia's Tiramisu	15	$17.99	$269.85
Jimmy's Crab House	Elkton	Frozen Alligator on a Stick	12	$27.99	$335.88
Jimmy's Crab House	Elkton	Brumley's Seafood Cocktail Sauce	24	$4.79	$114.96
Jimmy's Crab House	Elkton	Mama Mia's Tiramisu	18	$17.99	$323.82
Jimmy's Crab House	Elkton	Mom's Deep Dish Apple Pie	36	$12.49	$449.64

July 2011 Orders--Fine Foods, Inc.   Tuesday, August 09, 2011   11:04:41 AM

Total Orders $2,952.93

Page 1 of 1

FIGURE F-2 Access report

Also consider what kinds of graphic media are available and how well you can use them. Your employer will expect you to be able to use Microsoft PowerPoint to prepare your presentation as a slide show. Luckily, many college freshmen are required to take an introductory course that covers Microsoft Office and PowerPoint. If you are not familiar with PowerPoint, several excellent tutorials on the Web can help you learn the basics.

Anticipating the Unexpected

Even though you are only drafting your report and presentation at this stage, eventually you will answer questions from the audience. Being able to handle questions smoothly is the mark of a business professional. The first steps to addressing audience questions are being able to anticipate them and preparing your answers.

You will not use all the facts you gather for your report or presentation. However, as you draft your report, you might want to jot down those facts and keep them handy, in case you need them to answer questions from the audience. PowerPoint has a Notes section where you can include notes for each slide and print them to help you answer questions that arise during your presentation. You will learn how to print notes for your slides later in the tutorial.

The questions you receive depend on the nature of your presentation. For example, during a presentation of an Excel decision model, you might be asked why you are not recommending a certain course of action, or why you left it out of your report. If you have already prepared notes that anticipate such questions, you will probably remember your answers without even having to refer to the notes.

Another potential problem is determining how much technical detail you should display in your slides. In one sense, writing your report will be easier because you can include any graphics, tables, or data you want. Because you have a time limit for your presentation, the question of what to include or leave out becomes more challenging. One approach to this problem is to create more slides than you think you need, and then use the Hide Slide option in PowerPoint to "hide" the extra slides. For example, you might create slides that contain technical details you do not think you will have time to present. However, if you are asked for more details on a particular technical point, you can "unhide" a slide and display the detailed information needed to answer the question. You will learn more about the Hide Slide and Unhide Slide options later in the tutorial.

Writing the Introduction

After you have written the main body of your report and presentation, you can develop the introduction. The introduction should be only a paragraph or two, and it should preview the main points you will cover.

For some of the Access cases, you might want to include general information about databases: what they can do, why they are used, and how they can help a company become more efficient and profitable. You will not need to say much about the business operation because the audience already works for the company.

For the Excel cases, you might want to include an introduction of the general business scenario and describe any assumptions you used to create and run your decision support models. Excel is used for decision support, so you should describe the decision criteria you selected for the model.

Writing the Conclusion

Every good report or presentation needs a good ending. Do not leave the audience hanging. Your conclusion should be brief—only a paragraph or two—and it should give your presentation a sense of closure. Use the conclusion to repeat your main points or, for the Excel cases, to recap your findings and recommendations.

On many occasions, information learned during a business project reveals new opportunities for other projects. Your conclusion should provide closure for the immediate project, but if the project reveals possibilities for future improvements, include them in a "path forward" statement.

CREATING GRAPHICS

Visual aids are a powerful means of getting your point across and making it understandable to your audience. Visual aids come in a variety of forms, some of which are more effective than others. The integrated graphics tools in Microsoft Office can help you prepare a presentation with powerful impact.

Choosing Presentation Media

The media you use will depend on the situation and the media you have available, but remember: *You must maintain control of the media or you will lose the attention of your audience.*

The following list highlights the most common media used in a classroom or business conference room, along with their strengths and weaknesses:

- **PowerPoint slides and a projection system**—These are the predominant presentation media for academic and business use. You can use a portable screen and a simple projector hooked up to a PC, or you can use a full multimedia center. Also, although they are not yet universal in business, touch-sensitive projection screens (for example, Smart Board™ technology) are gaining popularity in college classrooms. The ability to project and display slides, video and sound clips, and live Web pages makes the projection system a powerful presentation tool. *Negatives:* Depending on the complexity of the equipment, you might have difficulties setting it up and getting it to work properly. Also, you often must darken the room to use the projector, and it may be difficult to refer to written notes during your presentation. When using presentation media, you must be able to access and load your PowerPoint file easily. Make sure your file is available from at least two sources that the equipment can access, such as a thumb drive, CD, DVD, or online folder. If your presentation has active links to Web pages, make sure that the presentation computer has Internet access.
- **Handouts**—You can create handouts of your presentation for the audience, which once was the norm for many business meetings. Handouts allow the audience to take notes on applicable slides. If the numbers on a screen are hard to read from the back of the room, your audience can refer to their handouts. With the growing emergence of "green" business practices, however, unnecessary paper use is being discouraged. Many businesses now require reports and presentation slides to be posted at a common site where the audience can access them later. Often, this site is a "public" drive on a business network. *Negatives:* Giving your audience reading material may distract their attention from your presentation. They could read your slides and possibly draw wrong conclusions from them before you have a chance to explain them.

- **Overhead transparencies**—Transparencies are rarely used anymore in business, but some academics prefer them, particularly if they have to write numbers, equations, or formulas on a display large enough for students to see from the back row in a lecture hall. *Negatives:* Transparencies require an overhead projector, and frequently their edges are visually distorted due to the design of the projector lens. You have to use special transparency sheets in a photocopier to create your slides. For both reasons, it is best to avoid using overheads.
- **Whiteboards**—Whiteboards are common in both the business conference room and the classroom. They are useful for posting questions or brainstorming, but you should not use one in your presentation. *Negatives:* You have to face away from your audience to use a whiteboard, and if you are not used to writing on one, it can be difficult to write text that is large enough and legible. Use whiteboards only to jot down questions or ideas that you will check on after the presentation is finished.
- **Flip charts**—Flip charts (also known as easel boards) are large pads of paper on a portable stand. They are used like whiteboards, except that you do not erase your work when the page is full—you flip over to a fresh sheet. Like whiteboards, flip charts are useful for capturing questions or ideas that you want to research after the presentation is finished. Flip charts have the same negatives as whiteboards. Their one advantage is that you can tear off the paper and take it with you when you leave.

Creating Graphs and Charts

Strictly speaking, charts and graphs are not the same thing, although many graphs are referred to as charts. Usually charts show relationships and graphs show change. However, Excel makes no distinction and calls both entities *charts*.

Charts are easy to create in Excel. Unfortunately, the process is so easy that people frequently create graphics that are meaningless, misleading, or inaccurate. This section explains how to select the most appropriate graphics.

You should use pie charts to display data that is related to a whole. For example, you might use a pie chart when breaking down manufacturing costs into Direct Materials, Direct Labor, and Manufacturing Overhead, as shown in Figure F-3. (Note that when you create a pie chart, Excel will convert the numbers you want to graph into percentages of 100.)

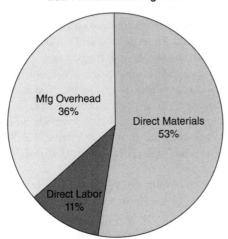

LCD TV Manufacturing Cost

FIGURE F-3   Pie chart: appropriate use

You would *not*, however, use a pie chart to display a company's sales over a three-year period. For example, the pie chart in Figure F-4 is meaningless because it is not useful to think of the period "as a whole" or the years as its "parts."

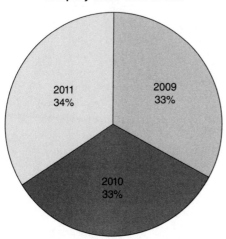

FIGURE F-4 Pie chart: inappropriate use

You should use vertical bar charts (also called column charts) to compare several amounts at the same time, or to compare the same data collected for successive periods of time. The same type of company sales data shown incorrectly in Figure F-4 can be compared correctly using a vertical bar chart (see Figure F-5).

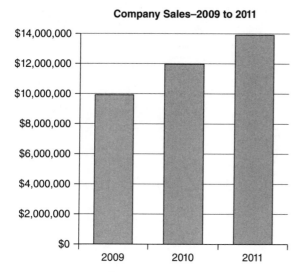

FIGURE F-5 Column chart: appropriate use

As another example, you might want to compare the sales revenues from several different products. You can use a clustered bar chart to show changes in each product's sales over time, as in Figure F-6. This type of bar chart is called a "clustered column" chart in Excel.

When building a chart, include labels that explain the graphics. For instance, when using a graph with an x- and y-axis, you should show what each axis represents so your audience does not puzzle over the graphic while you are speaking. Figures F-6 and F-7 illustrate the necessity of good labels.

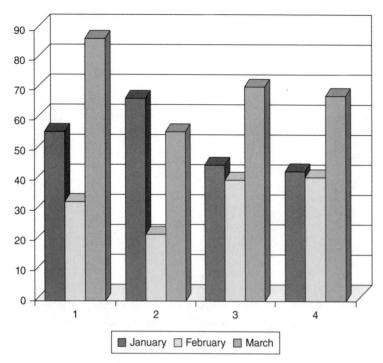

FIGURE F-6   3-D clustered column graph without labels

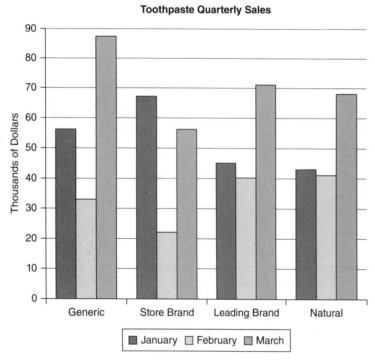

FIGURE F-7   Graph with labels

In Figure F-6, the graph has no title and neither axis is labeled. Are the amounts in units or dollars? What elements are represented by each cluster of bars? In contrast, Figure F-7 provides a comprehensive snapshot of product sales, which would support a talk rather than create confusion.

Another common pitfall of visual aids is charts that have a misleading premise. For example, suppose you want to show how sales are distributed among your inventory, and their contribution to net income. If you simply take the number of items sold in a given month, as displayed in Figure F-8, the visual fails to give your audience a sense of the actual dollar value of those sales. It is far more appropriate and informative to graph the net income for the items sold instead of the number of items sold. The graph in Figure F-9 provides a more accurate picture of which items contribute the most to net income.

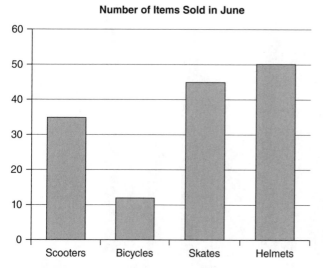

FIGURE F-8 Graph of number of items sold that does not reflect generated income

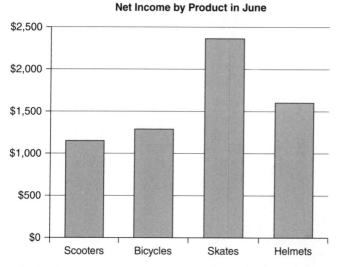

FIGURE F-9 Graph of net income by item sold

You should also avoid putting too much data in a single comparative chart. For example, assume that you want to compare monthly mortgage payments for two loan amounts with different interest rates and time frames. You have a spreadsheet that computes the payment data, as shown in Figure F-10.

	A	B	C	D	E	F	G
1	**Calculation of Monthly Payment**						
2	Rate	6.00%	6.10%	6.20%	6.30%	6.40%	6.50%
3	Amount	$ 100,000	$ 100,000	$ 100,000	$ 100,000	$ 100,000	$ 100,000
4	Payment (360 Payments)	$ 599	$ 605	$ 612	$ 618	$ 625	$ 632
5	Payment (180 Payments)	$ 843	$ 849	$ 854	$ 860	$ 865	$ 871
6	Amount	$ 150,000	$ 150,000	$ 150,000	$ 150,000	$ 150,000	$ 150,000
7	Payment (360 Payments)	$ 899	$ 908	$ 918	$ 928	$ 938	$ 948
8	Payment (180 Payments)	$ 1,265	$ 1,273	$ 1,282	$ 1,290	$ 1,298	$ 1,306

FIGURE F-10 Calculation of monthly payment

In Excel, it is possible (but not advisable) to capture all of the information in a single clustered column chart, as shown in Figure F-11.

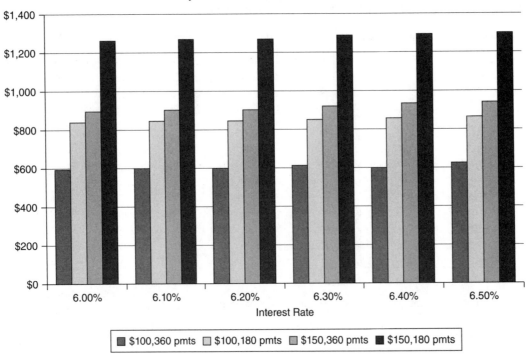

FIGURE F-11   Too much information in one chart

The chart contains a great deal of information. Putting the $100,000 and $150,000 loan payments in the same "cluster" may confuse the readers. They would probably find it easier to understand one chart that summarizes the $100,000 loan (see Figure F-12) and a second chart that covers the $150,000 loan.

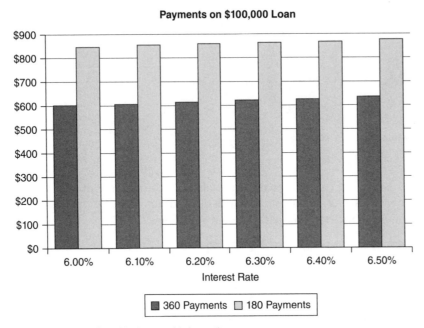

FIGURE F-12   Good balance of information

You could then augment the charts with text that summarizes the main differences between the payments for each loan amount. In that fashion, the reader is led step by step through the analysis.

Excel no longer has a Chart Wizard; instead, the Insert tab includes a Charts group. Once you create a chart and click it, three chart-specific tabs appear under a Chart Tools heading on the Ribbon to assist you with chart design, layout, and formatting. If you are unfamiliar with the charting tools in Excel, ask your instructor for guidance or refer to the many Excel tutorials on the Web.

Creating PowerPoint Presentations

PowerPoint presentations are easy to create. When you open PowerPoint, it starts a new presentation for you. You can select from many different themes, styles, and slide layouts by clicking the Design tab. If none of PowerPoint's default themes suit you, you can download theme "templates" from Microsoft Office Online. When choosing a theme and style for your slides, such as background colors or graphics, fonts, and fills, keep the following guidelines in mind:

- In older versions of PowerPoint, users were advised to avoid pastel backgrounds or theme colors and to keep their slide backgrounds dark. Because of the increasing quality of graphics in both computer hardware and projection systems, most of the default themes in PowerPoint will project well and be easy to read.
- If your projection screen is small or your presentation room is large, consider using boldface type for all of your text to make it readable from the back of the room. If you have time to visit the presentation site beforehand, bring your PowerPoint file, project a slide on the screen, and look at it from the back row of the audience area. If you can read the text, the font is large enough.
- Use transitions and animations to keep your presentation lively, but do not go overboard with them. Swirling letters and pinwheeling words can distract the audience from your presentation.
- It is an excellent idea to animate the text on your slides with entrance effects so that only one bullet point appears at a time when you click the mouse (or when you tap the screen using a touch-sensitive board). This approach prevents your audience from reading ahead of the bullet point being discussed and keeps their attention on you. Entrance effects can be incorporated and managed using the Add Animation button in PowerPoint 2010, as shown in Figures F-13 and F-14.

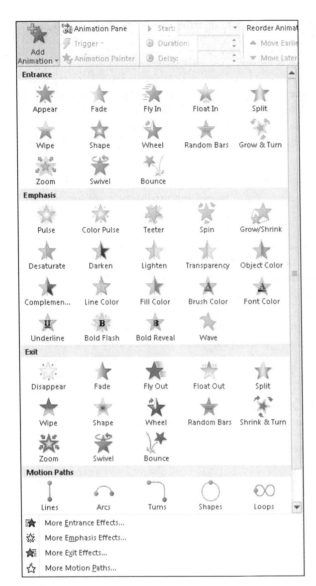

FIGURE F-13 The Add Animation button on the Ribbon in PowerPoint

FIGURE F-14 Add Entrance Effect window

N O T E—DIFFERENCES IN POWERPOINT ANIMATION TOOLS—2010 VS. 2007

The structure of the animation tools has changed considerably from PowerPoint 2007 to the 2010 version. The Custom Animation button and pane are both gone. Most of the custom animation tools are now incorporated using the Add Animation button in PowerPoint 2010. The look and feel is different, but the interface is more intuitive and easier to use. You can still use an animation pane to organize and edit your animations within a slide.

- Consider creating PowerPoint slides that have a section for your notes. You can print the notes from the Print dialog box by choosing Notes Pages from the Print menu, as shown in Figure F-15. Each slide will be printed at half its normal size, and your notes will appear beneath each slide, as shown in Figure F-16.

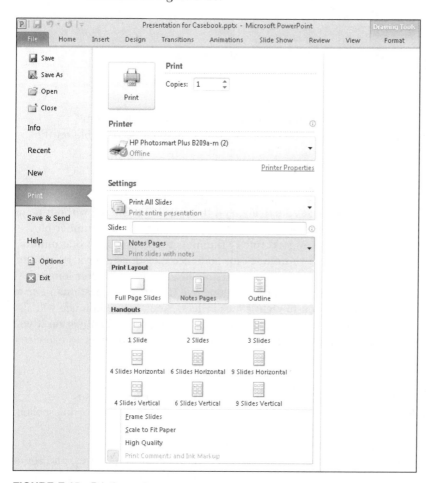

FIGURE F-15   Printing notes page

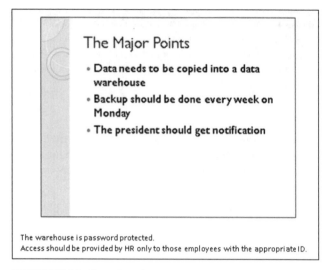

The warehouse is password protected.
Access should be provided by HR only to those employees with the appropriate ID.

FIGURE F-16 Sample notes page

- Finally, you should check your PowerPoint slides on a projection screen before your presentation. Information that looks good on a computer display may not be readable on the projection screen.

Using Visual Aids Effectively

Make sure you choose the visual aids that will work most effectively, and that you have enough without using too many. How many is too many? The amount of time you have to speak will determine the number of visual aids you should use, as will your target audience. A good rule of thumb is to allow at least one minute to present each PowerPoint slide. Leave a minimum of two minutes for audience questions after a 10-minute presentation, and allow up to 25 percent of your total presentation time to address questions after longer presentations. (For example, for a 20-minute presentation, figure on taking five minutes for questions.) For a 10-minute talk, try to keep the body of your presentation to eight slides or less. Your target audience will also influence your selection of visual aids. For instance, your slides will need more graphics and animation if you are addressing a group of teenagers than if you are presenting to a board of directors. Remember to use visual aids to emphasize your main points, not to detract from them.

Review each of your slides and visual aids to make sure it meets the following criteria:

- The font size of the text is large enough to read from the back of the presentation area.
- The slide or visual aid does not contain misleading graphics, typographical errors, or misspelled words—the quality of your work is a direct reflection on you.
- The content of your visual aid is relevant to the key points of your presentation.
- The slide or visual aid does not detract from your message. Your animations, pictures, and sound effects should support the text. Your visuals should look professional.
- A visual aid should look good in the presentation environment. If possible, rehearse your PowerPoint slides beforehand in the room where you will give the presentation. Make sure you can read your slides easily from the back row of seats in the room. If you have a friend who can sit in, ask her or him to listen to your voice from the back row of seats. If you have trouble projecting your voice clearly, consider using a microphone for your presentation.
- All numbers should be rounded unless decimals or pennies are crucial. For example, your company might only pay fractions of a cent per Web hit, but this cost may become significant after millions of Web hits.
- Slides should not look too busy or crowded. Many PowerPoint experts have a "6 by 6" rule for bullet points on a slide, which means you should include no more than six bullet points per slide and no more than six words per bullet point. Also avoid putting too many labels or pictures on a slide. Clip art can be "cutesy" and therefore has no place in a professional business presentation. A well-selected picture or two can add emphasis to the theme of a slide. For examples of a slide that is too busy versus one that conveys its points succinctly, see Figures F-17 and F-18.

Major Points

- Data needs to be copied into a data warehouse
- Backup should be done every week on Monday
- The president should get notification
- The vice president should get notification
- The data should be available on the Web
- Web access should be on a secure server
- HR sets passwords
- Only certain personnel in HR can set passwords
- Users need to show ID to obtain a password
- ID cards need to be the latest version

FIGURE F-17   Busy slide

The Major Points

- Data needs to be copied into a data warehouse
- Backup should be done every week on Monday
- The president should get notification

FIGURE F-18   Slide with appropriate number of bullet points and a supporting photo

You may find that you have created more slides than you have time to present, and you are unsure of which slides you should delete. Some may have data that an audience member might ask about. Fortunately, PowerPoint lets you "hide" slides; these hidden slides will not be displayed in Slide Show view unless you "unhide" them in Normal view. Hiding slides is an excellent way to keep detailed data handy in case your audience asks to see it. Figure F-19 shows how to hide a slide in a PowerPoint presentation. Right-click the slide you want to hide, and then click Hide Slide from the menu to mark the slide as hidden in the presentation. To unhide the slide, right-click it and then click Unhide Slide from the menu. Click the slide to display it in Slide Show view.

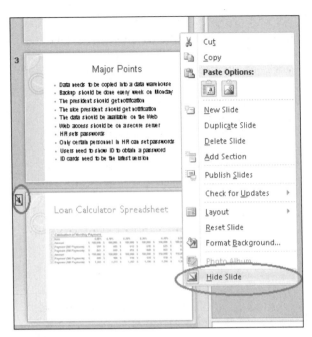

FIGURE F-19   Hiding a slide in PowerPoint

PRACTICING YOUR DELIVERY

Surveys indicate that public speaking is the greatest fear of many people. However, fear or nervousness can be channeled into positive energy to do a good job. Remember that an audience is not likely to think you are nervous unless you fidget or your voice cracks. Audience members want to hear what you have to say, so think about them and their interests—not about how you feel.

Your presentations for the cases in this textbook will occur in a classroom setting with 20 to 40 students. Ask yourself: Am I afraid when I talk to just one or two of my classmates? The answer is probably no. In addition, they will all have to give presentations as well. Think of your presentation as an extended conversation with several classmates. Let your gaze move from person to person, making brief eye contact with each of them randomly as you speak. As your focus moves from one person to another, think to yourself: I am speaking to one person at a time. As you become more proficient in speaking before a group, your gaze will move naturally among audience members.

Tips for Practicing Your Delivery

Giving an effective presentation is not the same as reading a report to an audience. You should rehearse your message well enough so that you can present it naturally and confidently, with your slides or other visual aids smoothly intermingled with your speaking. The following tips will help you hone the effectiveness of your delivery:

- Practice your presentation several times, and use your visual aids when you practice.
- Show your slides at the right time. Luckily, PowerPoint makes this easy; you can click the slide when you are ready to talk about it. Use cues as necessary in your speaker's notes.
- Maintain eye and voice contact with the audience when using the visual aid. Do not turn your back on your audience. It is acceptable to turn sideways to glance at your slide. A popular trick of experienced speakers is to walk around and steal a glance at the slide while moving to a new position.
- Refer to your visual aids in your talk and use hand gestures where appropriate. Do not ignore your own visual aid, but do not read it to your audience—they can read for themselves.
- Keep in mind that your slides or visual aids should support your presentation, not *be* the presentation. Do not try to crowd the slide with everything you plan to say. Use the slides to illustrate key points and statistics, and fill in the rest of the content with your talk.
- Check your time, especially when practicing. If you stay within the time limit when practicing, you will probably finish a minute or two early when you actually give the presentation. You will be a little nervous and will talk a little faster to a live audience.

- Use numbers effectively. When speaking, use rounded numbers; otherwise, you will sound like a computer. Also make numbers as meaningful as possible. For example, instead of saying "in 83 percent of cases," say "in five out of six cases."
- Do not extrapolate, speculate, or otherwise "reach" to interpret the output of statistical models. For example, suppose your Excel model has many input variables. You might be able to point out a trend, but often you cannot say with mathematical certainty that if a company employs the inputs in the same combination, it will get the same results.
- Some people prefer recording their presentation and playing it back to evaluate themselves. It is amazing how many people are shocked when they hear their recorded voice—and usually they are not pleased with it. In addition, you will hear every *um, uh, well, you know*, throat-clearing noise, and other verbal distraction in your speech. If you want feedback on your presentation, have a friend listen to it.
- If you use a pointer, be careful where you wave it. It is not a light saber, and you are not Luke Skywalker. Unless you absolutely have to use one to point at crucial data on a slide, leave the pointer home.

Handling Questions

Fielding questions from an audience can be tricky because you cannot anticipate all of the questions you might be asked. When answering questions from an audience, *treat everyone with courtesy and respect.* Use the following strategies to handle questions:

- Try to anticipate as many questions as possible, and prepare answers in advance. Remember that you can gather much of the information to prepare those answers while drafting your presentation. The Notes section under each slide in PowerPoint is a good place to write anticipated questions and your answers. Hidden slides can also contain the data you need to answer questions about important details.
- Mention at the beginning of your talk that you will take questions at the end of the presentation, which helps prevent questions from interrupting the flow and timing of your talk. In fact, many PowerPoint presentations end with a Questions slide. If someone tries to interrupt, say that you will be happy to answer the question when you are finished, or that the next graphic answers the question. Of course, this point does not apply to the company CEO—you *always* stop to answer the CEO's questions.
- When answering a question, a good practice is to repeat the question if you have any doubt that the entire audience heard it. Then deliver your answer to the whole audience, but make sure you close by looking directly at the person who asked the question.
- Strive to be informative, not persuasive. In other words, use facts to answer questions. For instance, if someone asks your opinion about a given outcome, you might show an Excel slide that displays the Solver's output; then you can use the data as the basis for answering the question. In that light, it is probably a good idea to have computer access to your Excel model or Access database if your presentation venue permits it, but avoid using either unless you absolutely need it.
- If you do not know the answer to a question, it is acceptable to say so, and it is certainly better than trying to fake the answer. For instance, if someone asks you the difference between the Simplex LP and GRG solving methods in Excel Solver, you might say, "That is an excellent question, but I really don't know the answer—let me research it and get back to you." Then follow up after the presentation by researching the answer and contacting the person who asked the question.
- Signal when you are finished. You might say that you have time for one more question. Wrap up the talk yourself and thank your audience for their attention.

Handling a "Problem" Audience

A "problem" audience or a heckler is every speaker's nightmare. Fortunately, this experience is rare in the classroom: Your audience will consist of classmates who also have to give presentations, and your instructor will be present to intervene in case of problems.

Heckling can be a common occurrence in the political arena, but it does not happen often in the business world. Most senior managers will not tolerate unprofessional conduct in a business meeting. However, fellow business associates might challenge you in what you perceive as a hostile manner. If so, remain calm, be professional, and rely on facts. The rest of the audience will watch to see how you react—if you behave professionally, you make the heckler appear unprofessional by comparison and you'll gain the empathy of the audience.

A more common problem is a question from an audience member who lacks technical expertise. For instance, suppose you explained how to enter data into an Access form, but someone did not understand your explanation. Ask the questioner what part of the explanation was confusing. If you can answer the question briefly and clearly, do so. If your answer turns into a time-consuming dialogue, offer to give the person a one-on-one explanation after the presentation.

Another common problem is receiving a question that you have already answered. The best solution is to give the answer again, as briefly as possible, using different words in case your original answer confused the person. If someone persists in asking questions that have obvious answers, you might ask the audience, "Who would like to answer that question?" The questioner should get the hint.

PRESENTATION TOOLKIT

You can use the form in Figure F-20 for preparation, the form in Figure F-21 for evaluation of Access presentations, and the form in Figure F-22 for evaluation of Excel presentations.

Preparation Checklist

Facilities and Equipment

☐ The room contains the equipment that I need.
☐ The equipment works and I've tested it with my visual aids.
☐ Outlets and electrical cords are available and sufficient.
☐ All the chairs are aligned so that everyone can see me and hear me.
☐ Everyone will be able to see my visual aids.
☐ The lights can be dimmed when/if needed.
☐ Sufficient light will be available so I can read my notes when the lights are dimmed.

Presentation Materials

☐ My notes are available, and I can read them while standing up.
☐ My visual aids are assembled in the order that I'll use them.
☐ A laser pointer or a wand will be available if needed.

Self

☐ I've practiced my delivery.
☐ I am comfortable with my presentation and visual aids.
☐ I am prepared to answer questions.
☐ I can dress appropriately for the situation.

FIGURE F-20   Preparation checklist

Evaluating Access Presentations

Course: _____ Speaker: _____   Date: _____

Rate the presentation by these criteria:
4=Outstanding   3=Good   2=Adequate   1=Needs Improvement
N/A=Not Applicable

Content

_____ The presentation contained a brief and effective introduction.

_____ Main ideas were easy to follow and understand.

_____ Explanation of database design was clear and logical.

_____ Explanation of using the form was easy to understand.

_____ Explanation of running the queries and their output was clear.

_____ Explanation of the report was clear, logical, and useful.

_____ Additional recommendations for database use were helpful.

_____ Visuals were appropriate for the audience and the task.

_____ Visuals were understandable, visible, and correct.

_____ The conclusion was satisfying and gave a sense of closure.

Delivery

_____ Was poised, confident, and in control of the audience

_____ Made eye contact

_____ Spoke clearly, distinctly, and naturally

_____ Avoided using slang and poor grammar

_____ Avoided distracting mannerisms

_____ Employed natural gestures

_____ Used visual aids with ease

_____ Was courteous and professional when answering questions

_____ Did not exceed time limit

Submitted by: _____

FIGURE F-21   Form for evaluation of Access presentations

Evaluating Excel Presentations

Course: _____ Speaker: _____   Date: _____

Rate the presentation by these criteria:
4=Outstanding   3=Good   2=Adequate   1=Needs Improvement
N/A=Not Applicable

Content

_____ The presentation contained a brief and effective introduction.

_____ The explanation of assumptions and goals was clear and logical.

_____ The explanation of software output was logically organized.

_____ The explanation of software output was thorough.

_____ Effective transitions linked main ideas.

_____ Solid facts supported final recommendations.

_____ Visuals were appropriate for the audience and the task.

_____ Visuals were understandable, visible, and correct.

_____ The conclusion was satisfying and gave a sense of closure.

Delivery

_____ Was poised, confident, and in control of the audience

_____ Made eye contact

_____ Spoke clearly, distinctly, and naturally

_____ Avoided using slang and poor grammar

_____ Avoided distracting mannerisms

_____ Employed natural gestures

_____ Used visual aids with ease

_____ Was courteous and professional when answering questions

_____ Did not exceed time limit

Submitted by: _____

FIGURE F-22 Form for evaluation of Excel presentations

INDEX